EXPLORING

Mammals

15

Pumas–Rats and mice

Marshall Cavendish
Reference
New York

Marshall Cavendish
99 White Plains Road
Tarrytown, NY 10591–9001

www.marshallcavendish.com

CONSULTANTS
John L. Gittleman, PhD, Director, Institute of Ecology, University of Georgia
James Martin, BSc, University of Bristol, UK
Nancy B. Simmons, PhD, Curator-in-Charge, Department of Mammalogy, American Museum of Natural History, New York
Erik Terdal, PhD, Associate Professor of Biology, Northeastern State University, Oklahoma

LIBRARY OF CONGRESS CATALOGING-IN-PUBLICATION DATA

Exploring mammals.
p. cm.
Includes bibliographical references and index.
ISBN 978-0-7614-7719-8 (set: alk. paper)
1. Mammals—Encyclopedias. I. Marshall Cavendish Corporation II. Title.

QL701.2.E97 2007
599.03—dc22

2007060864

ISBN 978-0-7614-7719-8 (Set)
ISBN 978-0-7614-7736-5 (Vol. 15)

PRINTED IN CHINA
11 10 09 08 07 1 2 3 4 5

MARSHALL CAVENDISH
Editor: Stephanie Driver
Publisher: Paul Bernabeo
Production Manager: Mike Esposito

THE BROWN REFERENCE GROUP PLC
Project Editor: Lesley Ellis
Copy Editors: Paul Thompson, Barbara Taylor, Sally McFall
Designers: Alison Gardner, Lynne Ross
Picture Researcher: Laila Torsun
Indexer: Kay Ollerenshaw
Cartographer: Darren Awuah
Design Manager: Sarah Williams
Managing Editor: Bridget Giles
Editorial Director: Lindsey Lowe

WRITERS
Amy-Jane Beer; Trevor Day; Leon Gray; Jen Green; Tom Jackson; Barbara Taylor

Photographic and illustration credits

All artworks and maps are copyright of Marshall Cavendish.

Front cover: **Shutterstock**. Back cover: **Shutterstock, Photos.com, Photos.com**.
Digital Stock: 1152b; **John Foxx Images:** 1139; **NHPA:** Adrian Hepworth 1166/1167; **NaturePL:** Dave Watts 1168, 1174/1175; **OSF:** Mike Powles 1138, Robin Redfern 1189; **Photodisc:** 1158/1159; **Photos.com:** 1125, 1152t; **Shutterstock:** 1153, EcoPrint 1182, Ronnie Howard 1124, 1131, Bruce MacQueen 1145, Steve MacWilliam 1183, Andy Stratiltov 1197, Gary Unwin 1169.

Contents

Pumas

Profile: The Largest Small Cat 1124
Anatomy: Puma 1128
Habitat and survival: From north to south 1130
Behavior: Nighttime loners 1132
Behavior: Pouncing on prey 1134
Behavior: Only meet to mate 1136

Rabbits

Profile: Breeding Success 1138
Anatomy: European rabbit 1142
Habitat and survival: Spread around the world 1144
Behavior: To dig or not to dig 1146
Behavior: Pellet power 1148
Behavior: Competition all the way 1150

Raccoons

Profile: Masked Bandits 1152
Anatomy: Common raccoon 1156
Habitat: From brushland to busy cities 1158
Behavior: Climbing to success 1160
Behavior: Flexible foragers 1162
Behavior: Spring into action 1164
Survival: Survivors of the fur trade 1166

Rat kangaroos

Profile: Tiny Hoppers 1168
Anatomy: Rufous rat kangaroo 1172
Habitat and survival: Outback animals 1174
Behavior: Silent night moves 1176
Behavior: Anything goes for omnivores 1178
Behavior: Pouch hangers-on 1180

Rats and mice

Profile: Gnawing Success 1182
Anatomy: Brown rat 1186
Habitat: All around the world 1188
Behavior: Sensing danger 1190
Behavior: Everything on the menu 1192
Behavior: Short, busy lives 1194
Survival: Island threats 1196

Mammal family tree 1198
Index 1200

PUMAS

Left: *A puma prepares to pounce on its prey, fixing its eyes on the target. At first glance, the puma looks a little like a lioness, but there are several differences. For example, unlike big cats, the puma lives alone and can purr but not roar.*

The Largest Small Cat

Less well known than their larger relatives—lions and tigers—pumas live across most of North and South America in a range of habitats. Biologists classify pumas as small cats, and they are the most widespread cat species in the world.

The continents of North and South America are home to 11 species of cats, including the jaguar, lynx, bobcat, ocelot, and puma. If a walker met a puma while hiking in the mountains, he or she would be forgiven for thinking it was a big cat and a close relative of lions, tigers, and other giant hunters. After all, pumas are sometimes known as mountain lions. However, only one American species of cat is classified as a big cat—the jaguar. All other American cats, including the puma, are classified as small cats. Therefore, pumas are really just a large relative of the house cat.

***Above:** Pumas have circular pupils, which are seen very clearly here. Like other cats, the puma is a hunter and has excellent eyesight and extremely sensitive hearing.*

KEY FACTS

- **Common names:** Puma, cougar, mountain lion, or panther
- **Scientific name:** *Puma concolor*
- **Subspecies:** Eastern cougar (*Puma concolor cougar*) and Florida cougar (*Puma concolor coryi*)
- **Family:** Felidae
- **Habitat:** Mountains, forests, prairies, including tropical forests, swamps, hilly forests, and snow-covered mountains
- **Range:** North and South America, from northern Canada to southern Argentina
- **Appearance:** Large cat with short, plain, gray-brown to black fur, which is very variable; small, round head; long, slender body; long, powerful legs; long tail with black tip.

Big small cat

The puma is also known as the cougar, mountain lion, or panther in some parts of its range. It is the world's largest small cat. Body size in pumas varies considerably across their range. There are very large pumas in southern South America, for example, which are probably heavier than small jaguars. However, there are several features that set apart the puma (subfamily Felinae) and the other 10 species of American cats from the jaguar (subfamily Pantherinae) and the other big cats.

One crucial difference is the hyoid bone, which is supported by muscles at the root of the tongue. The shape of the bone in small cats, including pumas, means they can purr. A big cat's hyoid bone lets it roar loudly.

> APART FROM THE HOUSE CAT, THE PUMA IS THE MOST WIDESPREAD OF ALL CAT SPECIES.

Also, if a puma curled up next to a house cat, both species of small cats would tuck their forefeet under the body when resting. However, big cats, such as lions and tigers, rest with their forelegs stretched out in front of them.

American cats

Apart from the house cat—a domestic breed of the Eurasian wild cat—which has been introduced to most parts of the world as a pet, the puma is the most widespread of all cat species. It lives across most of North and South America.

RELATIVES

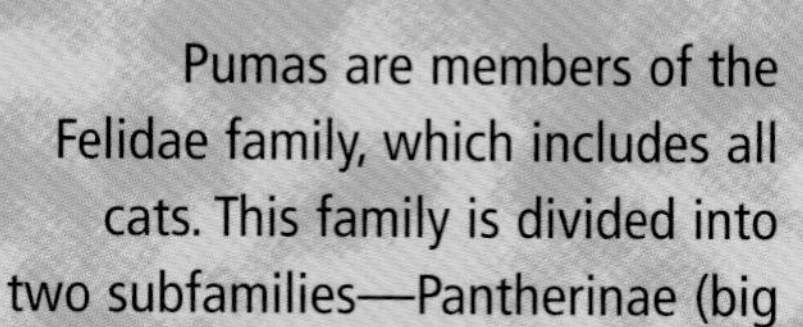

Pumas are members of the Felidae family, which includes all cats. This family is divided into two subfamilies—Pantherinae (big cats) and Felinae (small cats). Big cats include lions, tigers, jaguars, and leopards. Only one big cat—the jaguar—lives in the Americas. Most species of small cats live in Europe, Africa, and Asia. Pumas are one of the American small cats. Other American small cats include:

BOBCAT (*Lynx rufus*) ▶
Bobcats are sometimes called red lynx. These medium-sized cats live in rocky areas of North America, from southern Canada to southern Mexico.

CANADIAN LYNX (*Lynx canadensis*) ▶
The Canadian lynx lives farther north than any other American cat. It is a close relative of the northern lynx that lives in Europe and Asia.

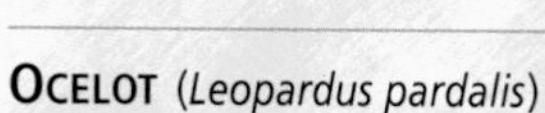

JAGUARUNDI (*Puma yaguarondi*) ▶
The closest relative of pumas, jaguarundis are small cats that live across Central and South America, as far south as Argentina. Jaguarundis are sometimes called otter cats because of their long body and short hair.

OCELOT (*Leopardus pardalis*)
The ocelot is another small cat that lives in the forests of Central and South America.

PAMPAS CAT (*Leopardus colocolo*)
As its named suggests, this cat lives in the pampas of South America. Pampas is the name of the South American prairies, which grow mainly in Argentina. This cat also lives in forests.

KODKOD (*Leopardus guigna*)
This cat is very rare. It is the smallest of the American cats and lives in the mountain forests of Chile and Argentina.

DID YOU KNOW?

Ancestors

The first cats evolved about 50 million years ago. At first these hunters did not look much like present-day cats. About 1 million years ago, species such as the mighty saber-toothed tiger first appeared. By this time, cats had evolved into two groups: the small, or feline, cats; and the big, or pantherine, cats, including the clouded leopard. Most cat species are feline, or small cats. Of the 30 small cat species, 11 live in North and South America. There are 6 big cat species, only one of which, the jaguar, lives in the Americas.

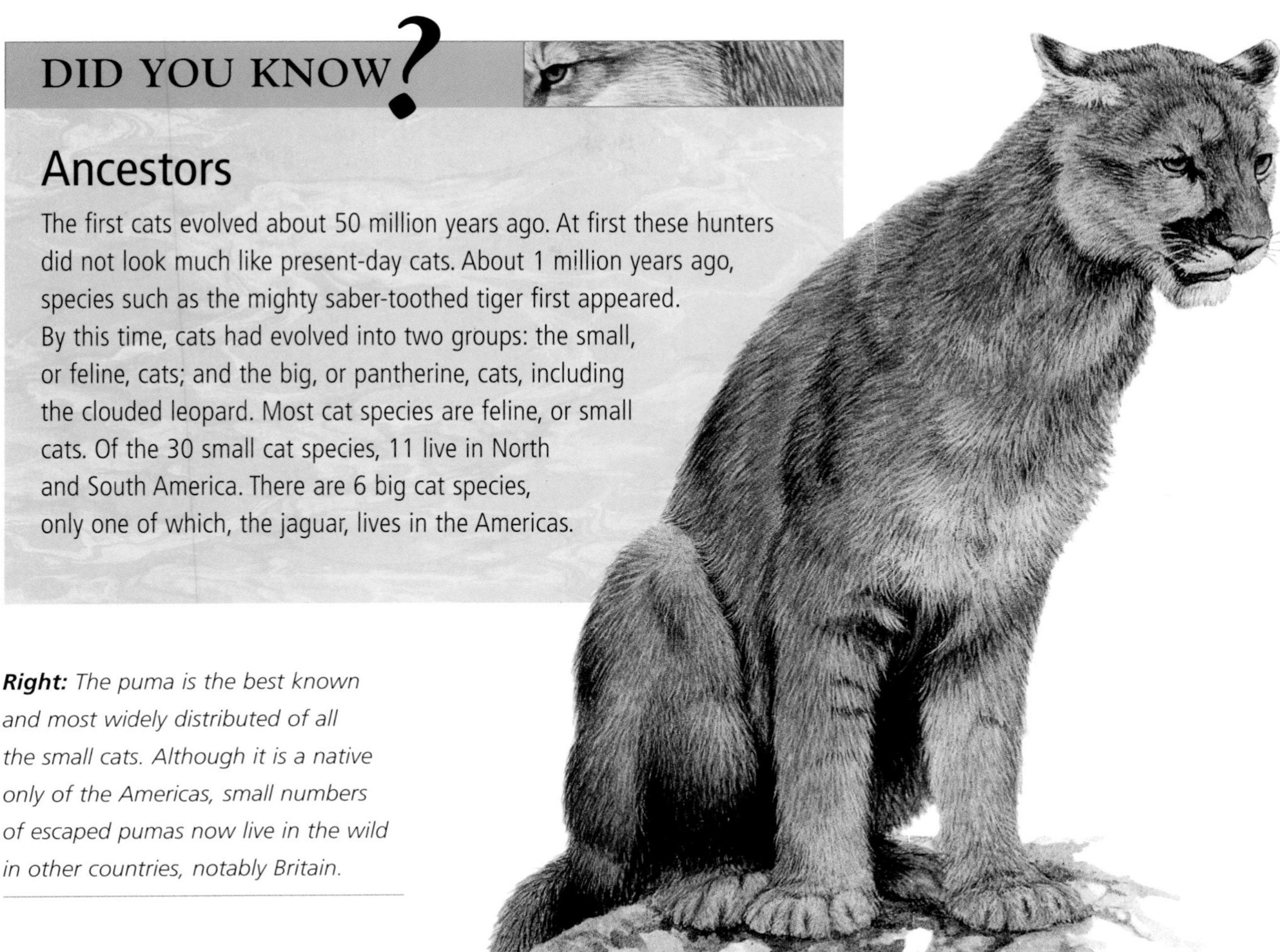

Right: *The puma is the best known and most widely distributed of all the small cats. Although it is a native only of the Americas, small numbers of escaped pumas now live in the wild in other countries, notably Britain.*

North America is also home to both the bobcat and the Canadian lynx. Most other American cat species live in Central and South America.

The ocelot and jaguarundi do live in the southwestern United States but they are now very rare in this area, if not extinct.

The margay, or tigrillo, lives in northern Mexico, but it is much more common farther south. The South American cat species are the tiger cat, Geoffroy's cat, pampas cat, mountain cat, and kodkod.

DID YOU KNOW?

Beasts on the loose

In many parts of the world there are local legends of giant, catlike killer animals on the loose. For example, many countryside areas of Britain, such as Exmoor and Bodmin, have their local "beast." There are regular sightings of large catlike animals and other evidence that shows that a beast is out there somewhere. Biologists believe that a few of these animals may be pumas or bobcats that were brought over from the Americas as pets. In 1976, a new law was passed in Britain to make sure that large and dangerous pets were kept properly. Many owners could not meet the demands of this new law and so released their cats into the wild. Since then, some of these cats have survived and even bred.

ANATOMY: Puma

Body shape
The puma's long and slender body is well suited to bursts of high-speed running and making long leaps.

Ears
The ears are very mobile so the puma can turn them toward the source of a sound.

Nose
Like all feline cats, the puma has a strip of naked skin across the top of its nose.

Mouth
Pumas do not smell just with their noses. As in other cats, they also have an odor-sensitive organ in the roof of their mouth. The odors enter the organ through two tiny holes.

Claws
The claws are retracted (pulled back) into the toes when the puma is walking. That prevents the claws from breaking or being blunted. When the puma pounces on a victim, tiny muscles pull the claws out ready for the attack.

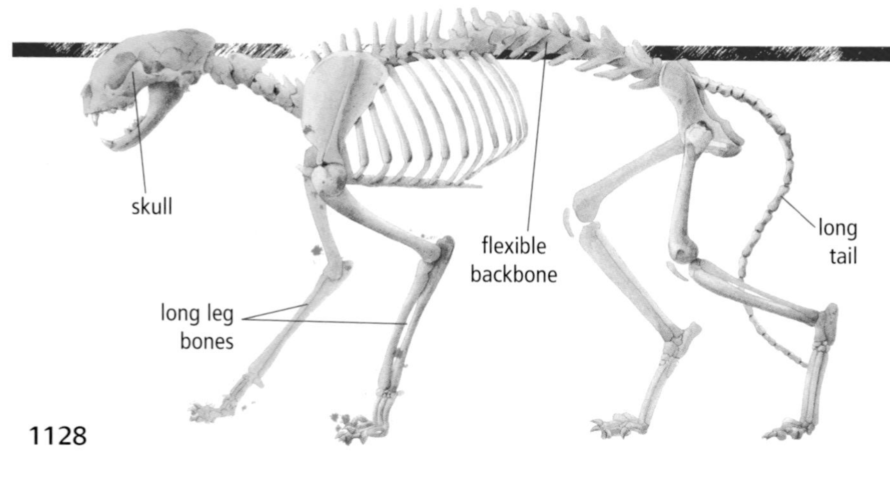

Skeleton
The shape of the puma's backbone supports the body but also allows great flexibility of movement, which is essential for a hunting cat that has to run and leap. The leg bones are very long and slender, and the skull is relatively small.

BLACK FUR

BLUE-GRAY FUR

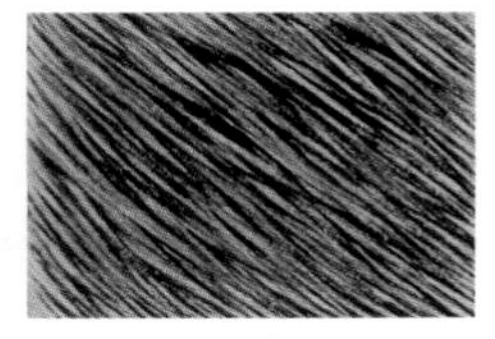

RED-BROWN FUR

Coat color

Pumas that live in different habitats usually have differently colored coats. Pumas living in tropical forests have reddish fur. Those living in colder areas, such as high up on mountains, have blue-gray fur. Most other pumas have pale brown fur. In a few rare cases, a puma may be completely black.

Tail

The puma's tail is long and tipped with black fur.

Legs

Like all cats, pumas stand on their toes. The long hand bones become part of the leg, making the limbs longer. Longer limbs are better for running and leaping.

FACT FILE

Pumas are the largest American cat species. Male pumas (above right) are larger than the females (above center). The next largest species is the lynx (above left), which is about 26–43 inches (67–110 cm) long. The smallest species is the kodkod, at just 20 inches (52 cm).

Puma

GENUS: *Puma*
SPECIES: *concolor*

SIZE

HEAD–BODY LENGTH: 41–79 inches (105–200 cm)
TAIL: 26–31 inches (67–78 cm)
WEIGHT: 79–231 pounds (36–105 kg)

COLORATION

Pumas have a coat with a single color, which varies from pale gray-brown to black.

Leg and toe bones

The long finger and toe bones fuse to become part of the leg. The claws stay inside sheaths in the feet (not visible here) when not in use.

claws

hand bones

FRONT LEG

HIND LEG

Skull

Compared with true big cats, such as lions and tigers, pumas have a small head and jaw.

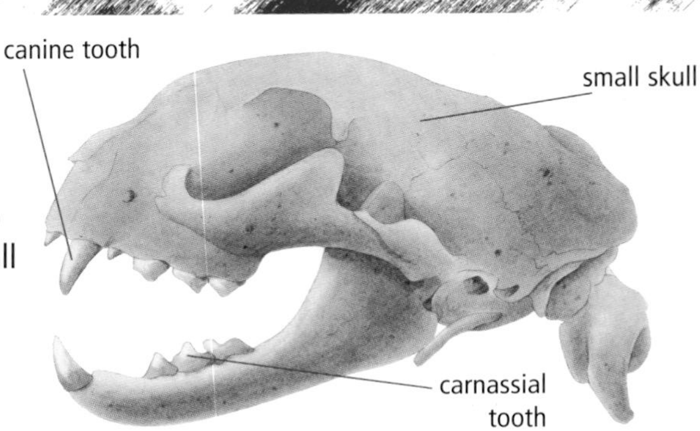

From north to south

The puma is the most widespread cat species in the world. It lives across most of North and South America and occupies a range of habitats. Many of the other American cats are more specialized.

The natural range of the puma spreads from the forests of southern Canada all the way through the United States and Central America, to Patagonia in southern Argentina. This area covers a variety of habitats, including everything from rocky mountain sides to deserts and rain forests. Now, however, hunting has forced pumas to remote areas.

Only the Canadian lynx lives farther north than the puma, in the cold conifer forests of Canada and the northern United States. In summer, the Canadian lynx also hunts on the tundra, the cold, treeless habitat that exists around the edge of the Arctic.

Bobcats are another species of small cat that lives in a range of habitats, including swamps, rocky areas, and mountain forests. This relative of the lynx lives across most of the United States, in parts of southern Canada, and as far south as southern Mexico. While the puma has been

THEN AND NOW

This map shows the current and former distribution of the puma.

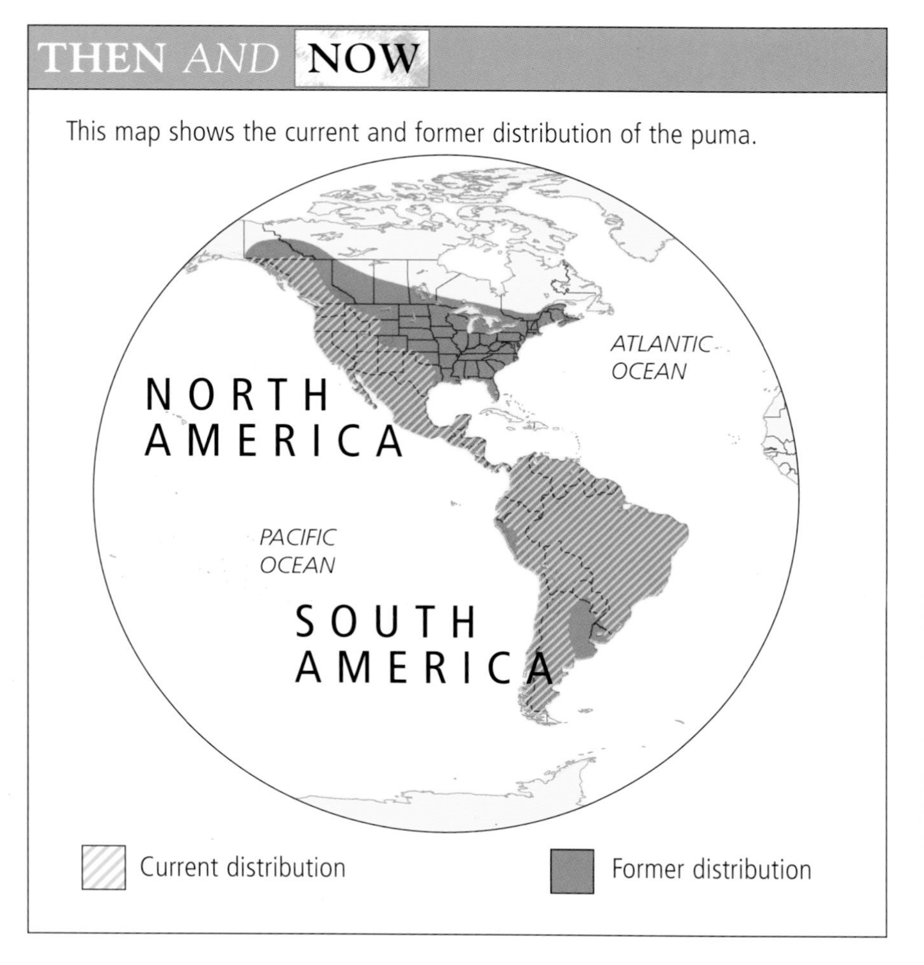

At risk

This chart shows how the International Union for the Conservation of Nature (IUCN) classifies pumas and other small cats:

FLORIDA PANTHER	*Critically endangered*
KODKOD	*Vulnerable*
MOUNTAIN CAT	*Endangered*

Critically endangered means that this species faces an extremely high risk of extinction in the wild. *Endangered* means that this species faces a very high risk of extinction. *Vulnerable* means that this species faces a high risk of extinction.

driven out of lowland areas of the United States by hunters, the smaller bobcat still survives across much of the country.

The ocelot and jaguarundi are two species that live in both North and South America. They can survive the desert conditions of northern Mexico and the southwestern United States, but both species are really forest cats and are expert climbers and swimmers.

Other forest cats include the kodkod, which lives in the mountains of Chile and Argentina; the tiger cat, which lives in rain forests; and the margay. The margay inhabits forests from Mexico to Argentina. It spends most of its time in trees and has extremely flexible feet that help it grip in steep, dangerous positions.

Geoffroy's cat and the mountain cat live in more open habitats at high altitudes.

DID YOU KNOW?

Puma country

A century ago, pumas lived across most of the warmer parts of North America. These cats hunted from coast to coast and reached as far north as the Yukon Territory of Canada. However, puma country now covers a much smaller area. As farmers took over more and more land, they regarded pumas as a danger to themselves and their animals. In reality, pumas rarely threaten people. Since 1900, pumas have attacked only 60 people in the whole of North America, just 9 of whom died as a result. Nevertheless, people often shot pumas on sight. As a result, pumas were forced to live in remote places where people did not go, such as mountains and deserts. These areas are where most pumas still live now.

Below: *A large puma stands in front of its mountainous habitat. Pumas are just as at home in mountainous regions as they are in swamps and deserts.*

Nighttime loners

Cats tend to be quiet animals that live on their own. The American species are no exception. Only the ocelot behaves differently because it sometimes lives in pairs. Other cats stay out of each other's way as much as possible.

Pumas and other American cats do not defend strict territories, but they do occupy a home range. The cats mark the boundary of their home ranges by rubbing their scent on landmarks, such as rocks or tree trunks.

A home range needs to be large enough to supply enough food for the cat. The size of the home range varies with the type of habitat. For example, a puma living in a tropical forest needs a smaller area to survive in than another puma that lives in a desert, where food is much more spread out.

PUMAS FOLLOW SET PATHS, WHICH ARE FREQUENTLY MARKED WITH SQUIRTS OF URINE AND CLAW MARKS.

In cold habitats there is a great deal of difference between winter and summer conditions. In these areas, pumas have a large winter range and a smaller summer range. Male pumas also have larger home ranges than females. That is also true for many other cat species. The male's range overlaps with those of several females.

Cats patrol their range to look for food. Pumas, for example, move around 25 miles (40 km) a day. They follow set paths, which are frequently marked with squirts of urine and claw marks on the ground.

Cats are most active at dusk or at night. They have excellent senses of smell and hearing. Cats are also known

AMAZING FACTS

- Tree-climbing margays have the most flexible hind feet of any cat. They can twist around by 180 degrees.
- Pumas can jump over obstacles that are around 6 feet (180 cm) high.
- According to the Native Americans of South America, jaguarundis were once kept in homes to control mice and rats.

Right: *The puma is active mainly at dusk and dawn and also at nighttime. That is when their prey is also on the move.*

Above: *A jaguarundi, or otter cat, drinks from a stream, remaining alert for danger. This small cat is a close relative of the puma.*

for seeing well in the dark. Like the eyes of many nocturnal mammals, cats' eyes have a reflective layer behind the light-sensitive retina. This mirrorlike layer reflects light that has not been picked up by the eye back onto the retina. As a result, the cats can see clearly in very low light. The cats' eyes also look very bright in the dark when a flashlight is shone on their face.

DID YOU KNOW?

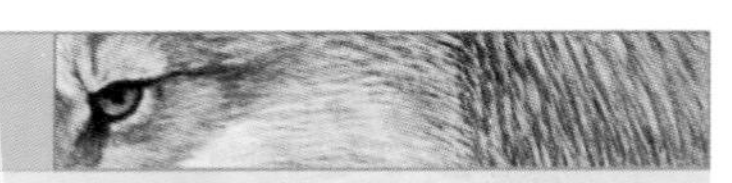

Highway help

Florida is home to a very rare subspecies of puma, called the Florida panther. These cats live in the Everglades, a huge wetland in southern Florida. They are also the state animal of Florida. In the 1960s, a highway was built right across the Everglades and threatened to cut the wetland in half. That would have meant the Florida panthers would have needed to cross the road to move around their large home ranges. Crossing the highway would have put the cats in grave danger of being run over. Therefore, engineers built tunnels under the highway, so now the panthers can safely get to the other side.

Pouncing on prey

All cats are hunters. Pumas and the other American cats have teeth and a digestive system suited to eating meat. People and other animals that eat nonmeat foods have flat cheek teeth that are good for grinding tough food. Cats do not have these flat teeth. Instead, the cusps (bumps) on a pair of teeth (one upper and one lower) have been modified to form a pair of blades. These teeth are called carnassials. As the upper and lower carnassials move past each other, they slice up mouthfuls of meat like scissors.

Above: *A bobcat has caught a large frog. As with other small cats, the size of a bobcat's territory usually depends on the availability of food in the area.*

PREY

Pumas and other cats are very efficient predators. They hunt alone and use their long canine teeth and powerful jaws to kill their prey. Cats kill a wide variety of prey, including:

DEER (family Cervidae)
Throughout the puma's range, its main prey is deer. The puma lies in wait and then pounces, delivering a fatal bite to the deer's neck.

RACCOON (*Procyon lotor*)
Small carnivores (meat eaters), such as raccoons, skunks, and weasels, are preyed on by cats.

BEAVER (*Castor canadensis*)
This large rodent lives in the forests of North America, where it is preyed on by lynx and bobcats.

MOUSE (family Muridae)
All over the world, small cats are known for hunting mice. Pet cats are kept to control the numbers of mice and other pests in buildings.

LIZARD (order Squamata)
In drier areas of the Americas, pumas and other cat species survive by hunting for lizards and other reptiles.

Below: *This puma has chased down and killed a deer. Pumas are strong enough to drag away their heavy prey and hide it from other meat eaters.*

DID YOU KNOW?

Feline teeth

A cat's chief weapons are its long fangs, which it uses to stab into victims. Once these teeth have grabbed the prey, it is very hard for a victim to shake free of the cat's powerful grip. Cats generally kill by biting the prey's neck. That action either snaps the prey's neck or crushes its windpipe so it cannot breath.

Compared with the size of its head, a cat's fangs are longer than those of any other carnivore, including dogs. However, biologists call these teeth *canines*, a word that also means "of a dog." Perhaps a cat's fangs should really be called "felines."

Above: *This rat will make a tasty meal for the margay. Margays spend more time in trees than any other small cat, dropping onto their prey from above.*

Meat is easy to digest, but it also contains high levels of nitrogen compounds. These substances are poisonous in large amounts, so meat eaters have to digest their food and get these compounds out of their body as quickly as possible. As a result, cats are typical carnivores (meat eaters) and have a short gut.

Cats catch prey by pouncing on them with their forepaws before providing the fatal bite. They tend to lie in wait or stalk a victim unseen before pouncing. Cats can run a short distance but rarely chase their prey far.

Larger cat species, such as the puma, lynx, and bobcat, can kill prey larger than themselves. Pumas favor hunting deer. A puma takes several days to eat a large deer.

Only meet to mate

The mating season is the only time of year when American cats allow another cat to get anywhere near them. The male and female puma can tell if the other is ready to mate by smelling one another's urine marks. Often the male sniffs out a mate in this way. However, the female might do the seeking by following the sounds of the male puma's loud mating calls.

A female cat's ovaries are stimulated to release eggs by the act of mating. That ensures that matings are generally successful. After mating, the male stays with the female for a few weeks to make sure that no other male breeds with his mate.

The female is ready to give birth after about 90 days. She finds a cave or shelter between rocks and lines the ground with leaves. The cubs are born blind. They can see from about the age of 10 days. They survive on milk for the first six weeks and then start to eat meat supplied by their mother. At three months, they begin a meat-only diet. The mother puma does not breed again until her cubs have grown up and left, probably at about the age of two.

SPOTTED AND RINGED: At birth the cubs have a spotted coat and a ringed tail. These markings will have faded when the cub reaches six months.

YOUNG ADULTS: Pumas become independent in their second year. Females will probably produce young in their third year, while males take a little longer to establish themselves.

HIDDEN AWAY: Cubs do not join the mother on hunting trips until they are about three months old. Before this age, they lie hidden in undergrowth while their mother is away finding food.

PLAYFUL KITTENS: Within weeks, baby pumas grow into strong, playful creatures and learn to climb trees.

The life of a puma

Puma

GESTATION: 3 months
LITTER SIZE: 1–6
WEIGHT AT BIRTH: 8–16 ounces (227–454 g)
WEANED: 3 months
INDEPENDENT FROM MOTHER: 1–2 years
SEXUAL MATURITY: Males 3 years; females 2.5 years
LIFE SPAN: 12 years; 20 years in captivity

Lynx

GESTATION: 2 months
LITTER SIZE: 2–3
WEIGHT AT BIRTH: 7 ounces (200 g)
FIRST SOLID FOOD: 1 month
WEANED: 5 months
INDEPENDENT FROM MOTHER: 10 months
SEXUAL MATURITY: Males 33 months; females 21 months
LIFE SPAN: Up to 15 years

LEARNING TO HUNT: Cubs are taught to hunt, including the tactics for hunting different prey, by their mother.

CHECK THESE OUT

RELATIVES: • Cheetahs • Jaguars • Leopards • Lions • Lynx and wildcats • Servals • Tigers
PREY: • Beavers • Deer • Raccoons • Rats and mice

RABBITS

With their long ears and powerful hind legs, rabbits are instantly recognizable. They are among the most successful animals on Earth and dig burrows in which to live and escape their many predators.

***Below:** A group of rabbits keeps a watchful lookout as they come out into the open to feed.*

Breeding Success

Rabbits are one of the world's most commonly seen wild mammals. While human activities have destroyed habitats and wiped out much of Earth's wildlife, many species of rabbits have thrived over the last 2,000 years or so. They have been so successful because of the changes people have made to the environment and to the rabbit's ability to breed very quickly. Farming activities, for example, have resulted in ideal habitats for rabbits, which eat most types of crops, including wheat and root vegetables.

Relative differences

Rabbits belong to a small order of mammals called lagomorphs. This group also includes hares and pikas.

Lagomorphs share many similarities with rodents, a much larger order of mammals, which includes rats, mice, beavers, and squirrels. Some types of rodents, such as the springhare, are often mistaken for rabbits. Like rodents, lagomorphs are small to

Above: *With its ears lying flat along its back, this European rabbit crouches low in the grass to avoid being seen. With so many predators, these rabbits rarely live longer than two years.*

KEY FACTS

- **Common name:** Rabbit
- **Species:** 26 species in 10 genera, including the European rabbit, eastern cottontail, and volcano rabbit
- **Family:** Leporidae
- **Habitat:** Wide ranging, including grasslands, woods, marshes, deserts, tropical rain forests, arctic tundra, swamps, mountain forests, and farmlands
- **Range:** Americas, Europe, Asia, Africa; introduced to Australia, New Zealand, and South America
- **Appearance:** Small brown animals with long ears and powerful hind legs. The fur is usually thick and soft and also covers the feet; shorter hair covers the ears. The tail is usually small and fluffy.

medium-sized mammals. Both types of mammals have self-sharpening, chisel-like teeth that are ideal for gnawing. The front teeth grow continuously and stay sharp as they wear against each other.

The ever-growing teeth of rodents and lagomorphs is a feature that both groups inherited from their most recent common ancestor. The genes of rabbits and rodents show that they evolved separately, but they are each other's closest relatives, and they live in similar ways.

Both lagomorphs and rodents have thickened enamel on the anterior (front) surfaces of their incisor teeth and may have thin enamel on at least part of the posterior (back) surfaces. Lagomorphs have two peglike teeth behind the upper incisors (front biting teeth).

DID YOU KNOW?

Ancestors

The first lagomorphs probably evolved around 55 million years ago. It is not known what these animals looked like, but they had peglike teeth just as present-day rabbits and hares have. About 38 to 35 million years ago, pikas began to evolve separately from other lagomorphs. That is why pikas look so different to rabbits and hares. The first pikas appeared in Asia around 30 million years ago.

The first rabbit evolved about 40 million years ago in North America. This species was called *Palaeolagus*, a name that means "ancient hare." This rabbit looked very similar to present-day species, but its hind legs were not as long. Most present-day rabbits and hares evolved from a single species that lived about 16 to 12 million years ago.

RELATIVES

Rabbits belong to an order of mammals called lagomorphs. Other lagomorphs include:

HARE (family Leporidae) ▶

Hares are the largest lagomorphs. They usually have longer legs than rabbits. Most hares belong to the genus *Lepus*. Most hares do not dig burrows. Instead, they rest in a shallow dip in the ground called a form. Hare species include the snowshoe hare of the Arctic and the brown hare.

JACKRABBIT (*Lepus* species)

Jackrabbits are a group of hares, most of which live in the western states of North America and parts of Mexico. Jackrabbits usually live in hot open areas. They have huge ears to help them lose heat and long legs for running quickly across open ground.

PIKA (family Ochotonidae) ▶

Pikas are smaller and look different from other lagomorphs. Pikas do not have long hind legs, and their ears are less pointed. Pikas live in cold habitats and have very thick fur. A pika's coat often makes the animal's body appear almost spherical.

All ears and legs

Lagomorphs are divided into two families. The 30 species of pikas belong to the family Ochotonidae, while rabbits and hares are in the family Leporidae. There are 62 species of rabbits and hares. These two types of lagomorphs are very similar to each other, with long ears that stick up above the head and powerful hind limbs that are much longer than the forelegs. There is no clear distinction between hares and rabbits, although hares are generally larger and have longer legs. Most hare species live out in the open and must be able to run away from predators. Rabbits are seldom far from the safety of a burrow or den and hop inside if a hunter appears. They do not need to run for so long and therefore have shorter legs.

Twenty-six members of the Leporidae family are rabbits. The most common and widespread of these is the European rabbit. Over the years, this species has spread across Europe and western Asia. It has also been introduced to South America, Australia, and several islands. All domestic rabbits are members of this species.

Other rabbit species include the cottontails. There are 17 species of cottontails. All of them live in North or South America. They are named for their fluffy white tails. The world's smallest rabbit, the pygmy rabbit, also lives in North America. The other rabbit species tend to be rare and are spread around other parts of the world.

Right: *The most obvious feature of the cottontail rabbit is its fluffy white tail.*

Above: *The European rabbit is very common and breeds at an incredible rate, producing between 10 and 30 offspring each year. All domesticated rabbits descend from this species.*

DID YOU KNOW?

Domestic rabbits

There are more than 40 domestic breeds of rabbits. Some types have been bred as pets, such as giant rabbits, which grow to more than 20 pounds (9 kg), and lop-eared rabbits, which have floppy ears. Other breeds, such as the angora, are bred for their long, soft coats.

ANATOMY: European rabbit

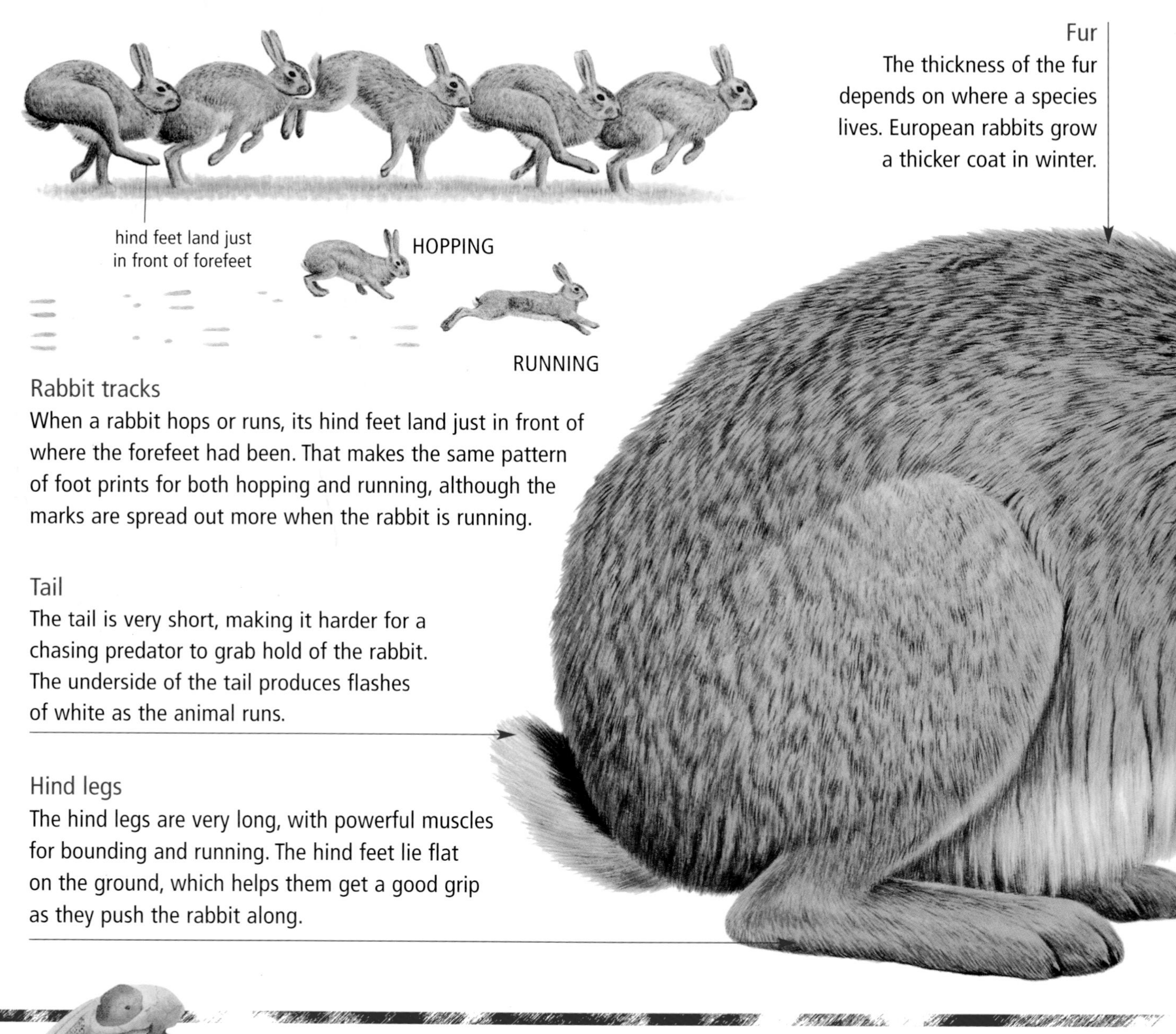

Fur

The thickness of the fur depends on where a species lives. European rabbits grow a thicker coat in winter.

Rabbit tracks

When a rabbit hops or runs, its hind feet land just in front of where the forefeet had been. That makes the same pattern of foot prints for both hopping and running, although the marks are spread out more when the rabbit is running.

Tail

The tail is very short, making it harder for a chasing predator to grab hold of the rabbit. The underside of the tail produces flashes of white as the animal runs.

Hind legs

The hind legs are very long, with powerful muscles for bounding and running. The hind feet lie flat on the ground, which helps them get a good grip as they push the rabbit along.

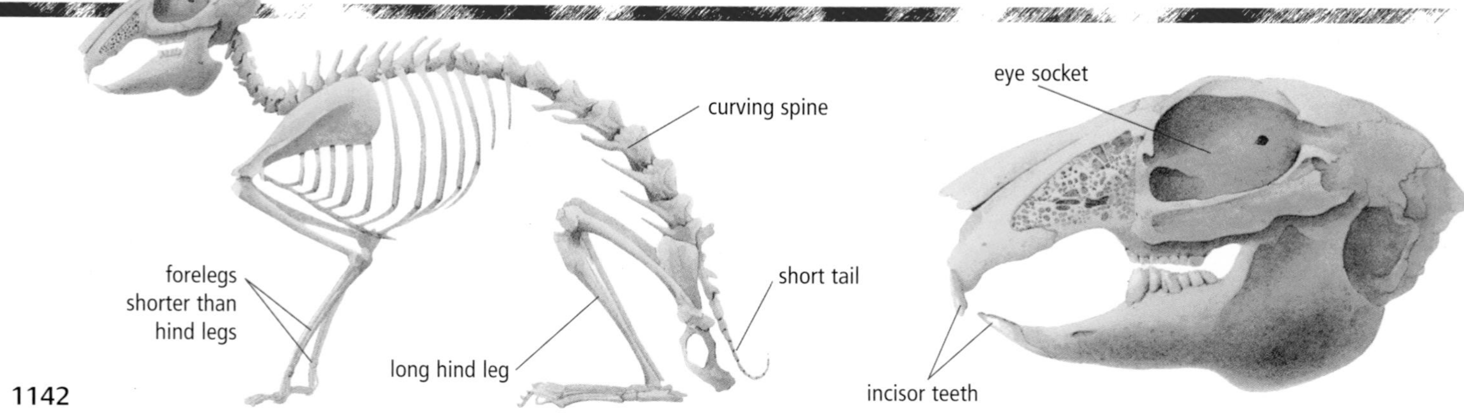

Ears
Hearing is the rabbit's most important sense. Its large ears can be swivelled around like a satellite dish to pick up sound from almost all directions. When the rabbit is running, its ears lie flat along the back.

Claws
Rabbits have long claws, which are sometimes used for digging burrows.

FACT FILE

Rabbits (center) are medium-sized lagomorphs. Hares (right) are larger, while pikas (left) are considerably smaller. The smallest rabbit is the pygmy rabbit, which is just 10 inches (25cm) long.

European rabbit

GENUS: *Oryctolagus*
SPECIES: *cuniculus*

SIZE

HEAD–BODY LENGTH: 20 inches (50 cm); females slightly smaller than males, with narrower heads
WEIGHT: 6.5 pounds (3 kg)

COLORATION

The coat is gray-brown. There are dark patches below the eyes. The tail is black on top and white underneath.

UNDERSIDE OF HARE'S SKULL

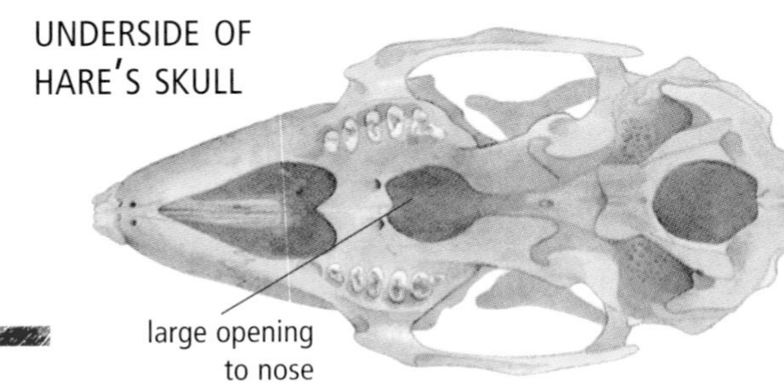

large opening to nose

UNDERSIDE OF RABBIT'S SKULL

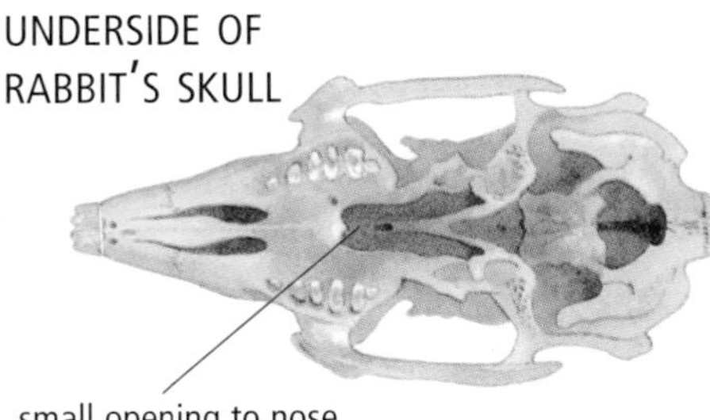

small opening to nose

Teeth
A rabbit's incisor teeth continue to grow throughout its life. The lower teeth grind down the upper teeth, keeping them very sharp and of the correct length, which is important for eating.

Skulls
Hares have larger air spaces inside their skull than rabbits. One reason may be that hares run over long distances, while rabbits generally just scurry a short way into cover. Hares may need to be able to breathe in more air than rabbits to keep going while on the run.

Spread around the world

Some rabbit species have evolved to survive in one type of habitat. For example, the volcano rabbit lives only in the grassy pine woodlands that grow on the slopes of just two volcanoes in central Mexico.

However, most species can survive in open habitats where there is plenty of grass and places to hide. Burrowing species also need areas where the soil is loose enough for them to dig in. A few rabbits live in different habitats. The desert cottontail survives in dry areas in the western United States, while the marsh rabbit, a swimming cottontail, lives in swamps in the southeastern United States.

Survivor

The most widespread rabbit species is the European rabbit. About 2,000 years ago, it probably lived only in Spain, Portugal, and parts of North Africa. The Romans then began to raise rabbits for eating and took them across their empire, from Britain to Turkey.

Some rabbits escaped and began to live wild, but most of Europe was too cold and wet for the rabbits to do well. However, 300 years ago Europe's land began changing into the system of farmed fields now in existence. This arrangement was ideal for the European rabbit, which began to steadily increase in numbers.

By the beginning of the twentieth century, rabbits had spread around the world and were becoming serious pests. Australia had the worst problem. In 1950 it was decided to infect these rabbits with a virus called myxomatosis. By 1953 the majority of the Australian rabbits had died.

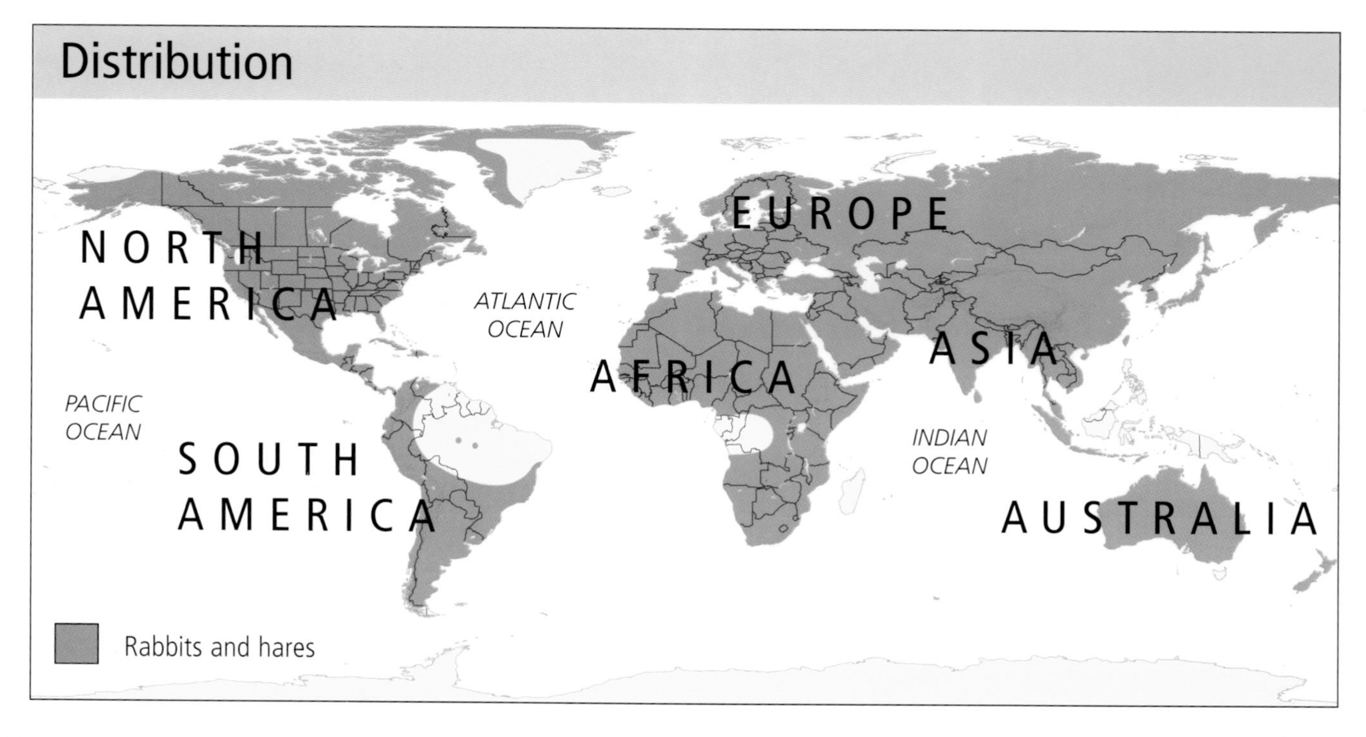

The myxomatosis virus spread to Europe, where it had the same effect. Since then, wild rabbits across the world have developed immunity to myxomatosis and their numbers have increased again.

Above: *A baby cottontail hides in the undergrowth near its burrow. Young rabbits stay close to the burrow so they can dart back to safety if a predator appears. Baby rabbits are ready to reproduce by the time they are just three months old.*

DID YOU KNOW?

On the up Down Under

Rabbits were introduced to Australia in 1859. A landowner named Austin Thomas released 24 rabbits on his land so he could hunt them. Rabbits do not have any natural predators in Australia, and Thomas's rabbits began to reproduce in huge numbers. Female rabbits can produce 36 babies, or kits, a year, and their female kits can breed when three months old. If all of Thomas's kits survived and bred as much as possible, then in just five years, the population of 24 rabbits could have grown to a staggering 2.6 trillion. Fortunately, the population did not grow quite so quickly. However, the actual size of the rabbit population had reached 600 million by the 1950s.

At risk

This chart shows how the International Union for the Conservation of Nature (IUCN) classifies rabbits:

New England cottontail	*Vulnerable*
Riverine rabbit	*Critically endangered*
Volcano rabbit	*Endangered*

Vulnerable means that this species faces a high risk of extinction in the wild. *Critically endangered* means that this species faces an extremely high risk of extinction in the wild and is likely to become extinct if nothing is done. *Endangered* means that this species faces a very high risk of extinction.

To dig or not to dig

Rabbits are most active at night, although sometimes they can be seen feeding during the day. When not feeding, rabbits rest out of sight. The European rabbit is well known for digging extensive networks of burrows, called warrens, where many rabbits live together.

However, this behavior is unusual for rabbits. Not all rabbit species are diggers. For example, none of the cottontail rabbits dig burrows. In areas where other burrowing animals live, cottontails often live in abandoned burrows. Otherwise they hide in thick undergrowth.

The few other rabbit species that do dig their own burrows, such as the pygmy rabbit and Bunyoro rabbit, also behave differently. Instead of living in social groups, or colonies, like the European rabbit, they spend most of their lives on their own.

Controlling space

Rabbits that dig burrows tend to control the area around them. That area is called their territory. A territory is generally only small, and rabbits always feed inside it.

The burrow is the safest place for rabbits, so they do not move too far away in case they need to dash underground to escape danger.

All the members of a warren defend the territory around their home. However, the dominant males in the group do most of the work. They also use their scent to mark the boundaries of the warren. This scent-spreading behavior ensures that dominant rabbits

AMAZING FACTS

- A rabbit's flashing white tail is easy to see as it runs; this means other rabbits can always see rabbits running from danger.
- The longest fence in the world was built in 1907 to keep rabbits out of Western Australia. The fence, called Number One Rabbit-Proof Fence, was 1,139 miles (1,837 km) long.

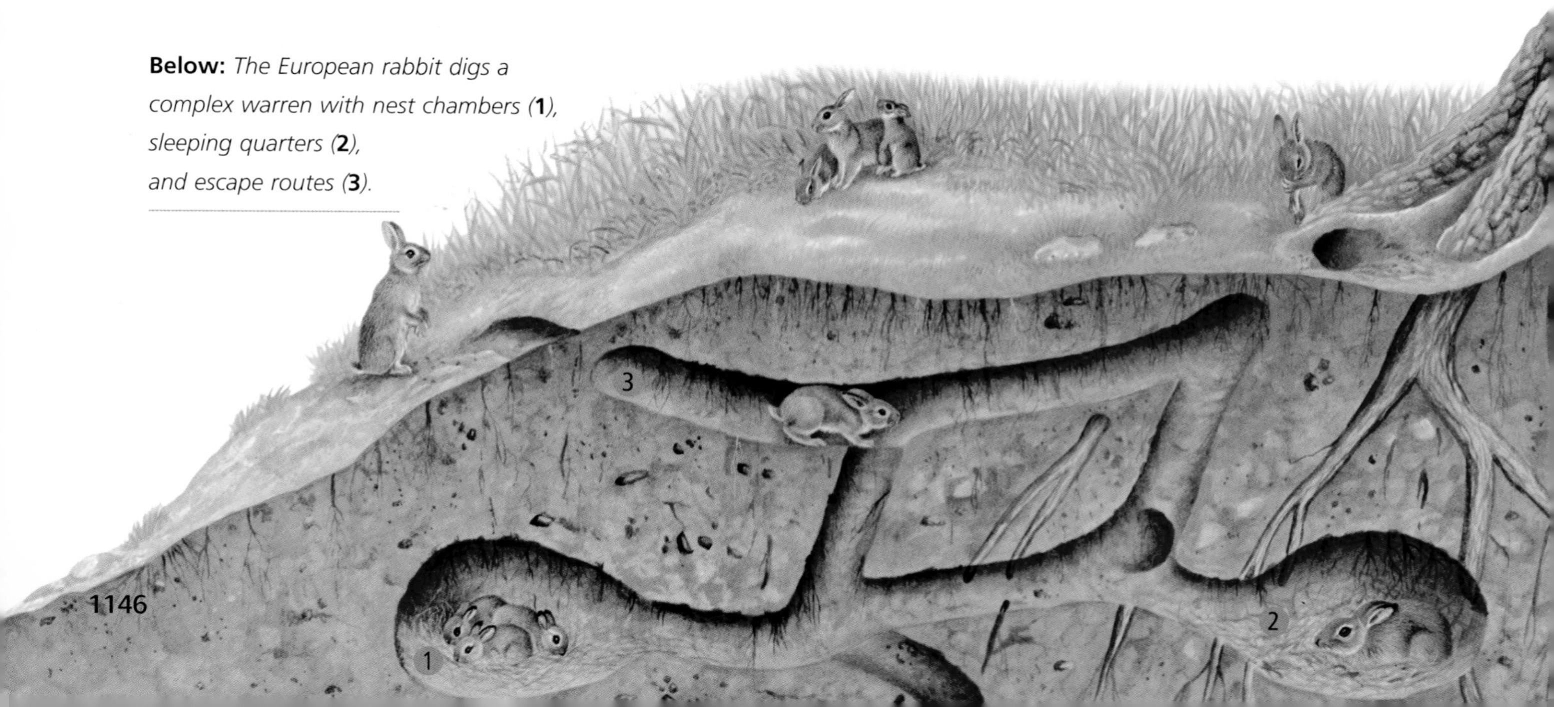

Below: *The European rabbit digs a complex warren with nest chambers (**1**), sleeping quarters (**2**), and escape routes (**3**).*

keep their important position among other rabbits in the warren.

Rabbits that do not have a burrow do not defend a territory. These species tend to move around a great deal between resting sites and are not based in any particular place.

Scents and scrapes

Rabbits mark their territory in a number of ways. They leave scrapes on the ground with their claws as a visual signal. Scent markings are also very important. There is a scent gland under each rabbit's chin. A rabbit wipes this gland on plants in a practice called chinning. There is also a gland next to the rabbit's anus, which mixes the animal's scent into its droppings. Rabbits usually produce droppings in obvious places around their territory.

Above: *Some species of rabbits dig new burrows in soft soil. They usually space themselves out and have large home ranges. Where the soil is hard and difficult to dig, however, they use the same burrows all the time, and their home ranges overlap. That leads to constant disagreements over territory.*

Pellet power

Rabbits eat mainly plants, although they do sometimes eat snails and earthworms. A rabbit's diet is usually made up of grass and the leaves and stems of other small plants. In winter, rabbits strip off the bark from trees when there is no other food to be found.

Plant food is very hard to eat. For example, grass is covered in tiny specks of silica, a sandlike substance. Silica wears down the teeth of animals that eat grass. The teeth of old sheep and cattle may be so worn that the animals can no longer chew.

DIGESTED FOOD COMES OUT OF THE RABBIT'S ANUS AS SLIMY PELLETS.

A rabbit's teeth also wear down. However, they keep growing throughout the animal's life, so they never become blunt.

Plants, made of tough fibers of cellulose, are also hard to digest. Cellulose is full of sugar but it does not break down inside a rabbit's stomach. Like other plant eaters, rabbits have bacteria (single-celled microorganisms) in their large intestine that break down the cellulose for them.

However, the rabbit cannot absorb the nutrients released by the bacteria into its blood from the large intestine. Instead, the digested food comes out of the rabbit's anus as soft pellets, which the rabbit then eats. That behavior gives the rabbit a second chance to extract nutrients from the food. After the second digestion, the waste comes out as dry pellets.

PREDATORS

Rabbits are favorite prey for many predators, including coyotes, lynx, and large birds of prey. A rabbit is lucky to survive for more than a year before it is killed by a hunter or by disease. Other rabbit predators include:

Fox (family Canidae)
All species of foxes hunt rabbits and hares.

Ermine, or stoat (family Mustelidae)
Ermine and other similar carnivores, such as polecats, have a long, narrow body. That makes them ideally suited for chasing rabbits down into their burrows and killing them underground.

Bobcat (*Lynx rufus*)
The bobcat is one of the larger species of wildcats that lives in North America. Bobcats hunt cottontail rabbits and hares in a range of habitats.

DID YOU KNOW?

Panoramic view

A rabbit's eyes are positioned on either side of the head, so each eye is looking more or less in the opposite direction. That arrangement allows the rabbit to see almost all the way around itself. The only place it cannot see is directly behind. When an animal's eyes are on the front of the head, pointing forward, they are both looking at the same scene. The brain uses this overlap to produce a detailed image of the surroundings. The views from each rabbit eye do not overlap, so the rabbit sees only a simple flat image. However, that is enough for the rabbit to detect the movements of an approaching predator, so it has time to run for cover.

Below: *This diagram shows a rabbit's "double digestion." First, the partly digested, nutrient-rich food comes out of the anus as small, soft pellets, which the rabbit eats. These pellets are digested again, and the waste emerges as large, hard droppings.*

Above: *A pair of young rabbits forages for food. Baby rabbits are weaned by the age of just three or four weeks, after which they are able to look for their own food.*

Competition all the way

European rabbits live in an ordered society. A male rabbit, or a buck, tends to dominate the females, or does, that live near his burrow. He mates with them frequently and makes sure that other bucks do not come near them. A doe is able to get pregnant for around one day a week.

During the breeding season, a buck's dominance is frequently tested by other males, and clashes often result. These clashes generally result in spectacular contests, where the bucks leap high into the air and attempt to slash their opponent with their claws.

Females also compete with each other for the best place in a burrow to give birth. Low-ranking does must give birth in dead-end tunnels, or stops. With only one way in or out, rabbits born in stops are more vulnerable to attack.

The doe lines the nursery burrow with dry leaves a few days before she gives birth. Baby rabbits are born helpless. They stay in the burrow or nest for several days before they are strong enough to come out into the open. The babies stop drinking their mother's milk after around a month. That allows the mother's body to prepare to mate again. By the age of three months, does are sexually mature. Young bucks are not fully grown until five months.

SPRAY AWAY: Before mating, a buck sprays a doe with urine. That gives the female rabbit his scent and warns other bucks to stay away from her. A doe is ready to mate again a month after giving birth.

European rabbit

MATING SEASON: Late winter and spring
GESTATION: 30 days
LITTER SIZE: 3–7
WEIGHT AT BIRTH: 1.4 ounces (40 g)
EYES OPEN: 10 days
WEANED: 4 weeks
SEXUAL MATURITY: Males 5 months; females 3 months
LIFE SPAN: 1–2 years

DID YOU KNOW?

Efficient breeders

A rabbit's life is often very short. However, rabbits reproduce very quickly so they make the most of what time they have. Rabbits have an efficient breeding system. For example, if a pregnant doe is having difficulty finding food or water, her body reabsorbs the embryos growing inside her. That action saves her wasting energy giving birth to offspring at a time when they are unlikely to survive.

CHECK THESE OUT

RELATIVES: • Hares and pikas
PREDATORS: • Coyotes • Dogs • Foxes • Jackals • Lynx and wildcats • Marten • Weasels

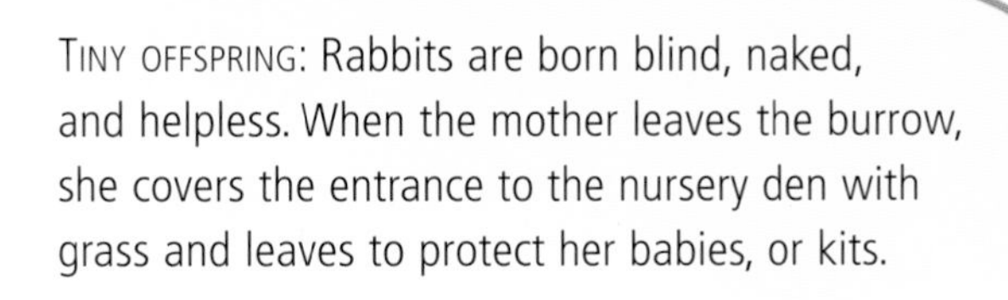

TINY OFFSPRING: Rabbits are born blind, naked, and helpless. When the mother leaves the burrow, she covers the entrance to the nursery den with grass and leaves to protect her babies, or kits.

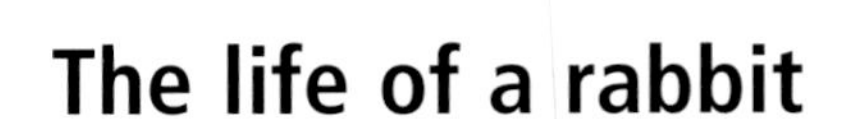

The life of a rabbit

CARRY OUT: Does carry their kits by the scruff of the neck.

RACCOONS

With its black face mask, ringed tail, and cheeky habits, the common raccoon is well known across much of North America. It is equally at home in the middle of a bustling city and in a quiet woodland. As a result, most Americans come across a raccoon from time to time, most probably going through their garbage can.

Above: *With its masklike facial markings, it is no surprise that the red panda was once thought to be a relative of the raccoons. However, biologists have now decided that it belongs to its own separate family, Ailuridae.*

Left: *A common raccoon balances on a log as it looks for food alongside a river. Raccoons eat a wide range of food, including fish, crayfish, clams, snails, fruit, vegetables, insects, and earthworms.*

Masked Bandits

The common raccoon is just one member of a family of animals called Procyonidae. This raccoon family also includes several species of less common raccoons, which live in South America and in the Caribbean, as well as the ringtail and cacomistle, olingos, coatis, and the kinkajou.

Raccoons and their relatives are called procyonids. This name comes from *pro*, the ancient Greek prefix for "before," and *cyon*, which means "dog." Early biologists thought that the procyonids were perhaps the ancestors of dogs. One species of wild dog looks so much like the procyonids that it is called the raccoon dog. However, raccoons and dogs evolved separately within the order Carnivora.

All procyonids are relatively small animals. They have a long body and a long, often bushy, and ringed tail. Most species have five toes on each foot and long claws that cannot be drawn into the foot.

Raccoons are the largest members of the family, followed by coatis, the kinkajou, and olingos. The smallest members are the ringtail and the less familiar cacomistle.

The earliest fossils of members of the raccoon family that have been found are 20 million years old. They were most similar to the modern cacomistle and ringtail species.

Meet the family

There are three raccoon species. The two main species are the common

KEY FACTS

- **Common name:** Common raccoon
- **Scientific name:** *Procyon lotor*
- **Family:** Procyonidae
- **Habitat:** Woodlands and bushy areas
- **Range:** Southern Canada, United States, Mexico, and Central America
- **Appearance:** The common raccoon has black, masklike stripes across the face and a ringed tail. The coat is usually grizzled gray; the ears are short, rounded, and tipped with light-colored fur.

Above: *The coati uses its long, flexible snout to forage among the leaves on the forest floor to find insects, lizards, and spiders.*

RELATIVES

Raccoons belong to the Procyonidae family of mammals. This family is one of many in the order of mammals called Carnivora. This order includes many hunting animals, including:

CAT (family Felidae) ▶
The cat family includes all types of cats, from the house cat to the giant tiger and lion. Cats live all over the world.

DOG (family Canidae) ▶
The dog family contains wolves, foxes, jackals, coyotes, and hunting dogs. Apart from foxes, which hunt alone, most dog species live in large groups called packs.

BEAR (family Ursidae) ▶
Bears are the largest carnivores (meat eaters). Despite their huge size and strength, bears are not active hunters. They do eat meat, but their diet also includes fruit and mushrooms. One bear species, the giant panda, was once thought to be a huge member of the raccoon family, Procyonidae.

MONGOOSE (family Herpestidae) ▶
These small carnivores live in Africa and warm parts of Asia. Mongooses have a powerful cylindrical body. Most of these animals live in groups.

OTTER, WEASEL, AND BADGER (family Mustelidae) ▶
The mustelids are a family of small carnivores that includes otters, weasels, and badgers. Most mustelids have a long flexible body. Mustelids live all over the world.

raccoon and crab-eating raccoon. The third is the Cozumel Island raccoon.

Raccoons are known for their distinctive face markings, which are not present in other family members. However, most of the other procyonids have the same bushy, ringed tails. As their name suggests, ringtails have characteristic rings on the tail, as does the closely related cacomistle.

RACCOONS ARE KNOWN FOR THEIR DISTINCTIVE FACE MARKINGS, WHICH ARE NOT PRESENT IN OTHER FAMILY MEMBERS.

The exception is the kinkajou, which lives in the forests of South and Central America. This animal has pale brown fur all over its body. The kinkajou's tail is unringed. It is also prehensile; it can wrap around branches and grip on like a fifth limb. The five olingo species are very similar to kinkajous, but they have pale rings on their tails. Their tails are not prehensile.

Right: *The kinkajou has extremely nimble forepaws, which help it climb among the trees and hold its food. It has a prehensile, or gripping, tail, which it wraps around branches while it swings through the trees.*

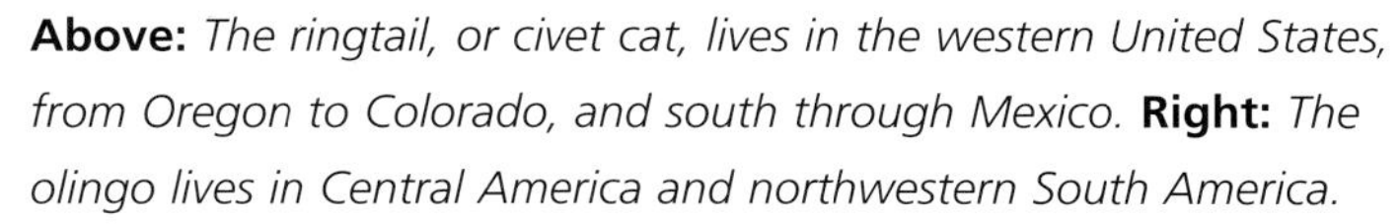

Above: *The ringtail, or civet cat, lives in the western United States, from Oregon to Colorado, and south through Mexico.* **Right:** *The olingo lives in Central America and northwestern South America.*

The four species of coatis are different from raccoons in other ways. They have ringed tails but shorter hairs than raccoons. As a result, the coati's body is less rounded, and the tail is less bushy. Coatis are known for their long and highly flexible snout, which sticks out in front of the jaw and is used for sniffing out and feeling for prey.

Below: *The common raccoon has distinctive masklike face markings.* **Below right:** *Coatis have a long, upturned, and very mobile snout.*

DID YOU KNOW?

Red panda—raccoon or bear?

The red panda was once placed in the raccoon family, Procyonidae, and it has also been connected to the bear family, Ursidae. However, most biologists now place the red panda on its own in the family Ailuridae. At one time, the red panda was thought to be a close relative of the giant panda, which is now placed in the bear family Ursidae. Both species live in the mountains of China, where they eat bamboo, and both have an unusual fingerlike projection on each forepaw, which they use to strip and eat bamboo. However, red pandas have a ringed tail like a raccoon, and giant pandas have black markings on the face similar to those of raccoons. Biologists were not sure whether the giant panda was a large raccoon or whether the red panda was a tiny bear. However, the two species' genes show that the two are not closely related. The giant panda is definitely a bear, while it is still not clear whether red pandas are more closely related to raccoons or to bears.

ANATOMY: Common raccoon

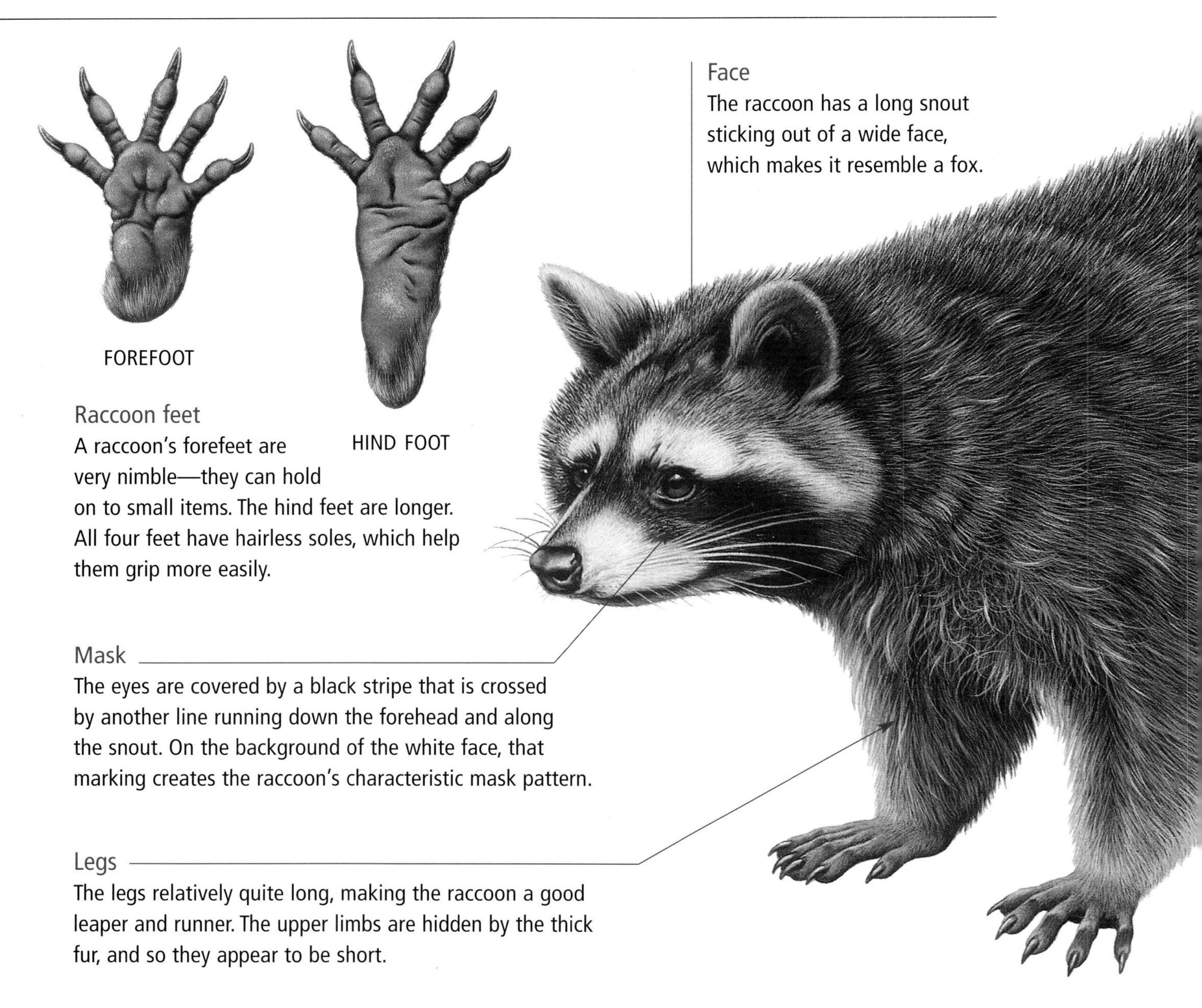

Face

The raccoon has a long snout sticking out of a wide face, which makes it resemble a fox.

Raccoon feet

A raccoon's forefeet are very nimble—they can hold on to small items. The hind feet are longer. All four feet have hairless soles, which help them grip more easily.

Mask

The eyes are covered by a black stripe that is crossed by another line running down the forehead and along the snout. On the background of the white face, that marking creates the raccoon's characteristic mask pattern.

Legs

The legs relatively quite long, making the raccoon a good leaper and runner. The upper limbs are hidden by the thick fur, and so they appear to be short.

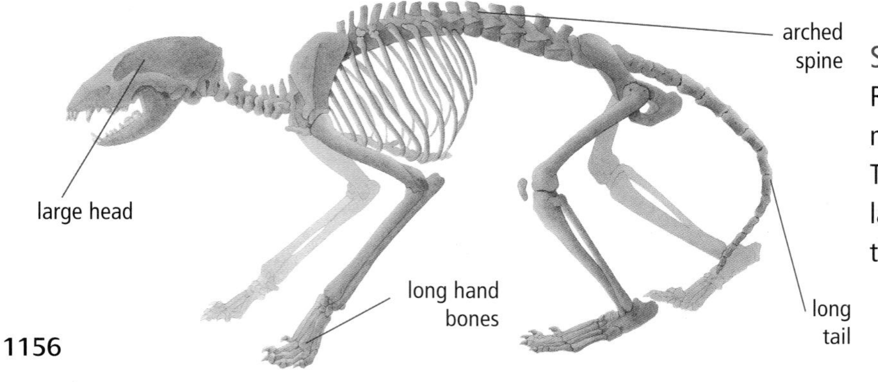

Skeleton

Raccoons have an arched spine, which makes its back the tallest point in its body. The arch helps counterbalance the raccoon's large head, which is often held close to the ground as the animal looks for food.

Fur
The coat has two layers of hairs. Soft and woolly hairs form a warm undercoat. Longer and thicker hairs form the outer layer. The thick fur makes the raccoon's body look larger than it really is.

Tail
The bushy tail has between 5 and 10 black or brown rings. The tip is always dark.

FACT FILE

The common raccoon (above left) is about the size of a medium-sized dog. The ring-tailed coati (above center) is 32 to 50 inches (81–127 cm) from the nose to tail tip. The kinkajou (above right) is 32 to 44 inches (81–113 cm) from nose to tail.

Common raccoon

GENUS: *Procyon*
SPECIES: *lotor*

SIZE

HEAD–BODY LENGTH: 16–37 inches (41–94 cm)
TAIL LENGTH: 7.5–16 inches (19–41 cm)
WEIGHT: 11–46 pounds (5–21 kg); males are larger than females

COLORATION

The coat is grizzled gray and made up of dark hairs with pale tips. The tail has black or dark brown rings and a dark tip.

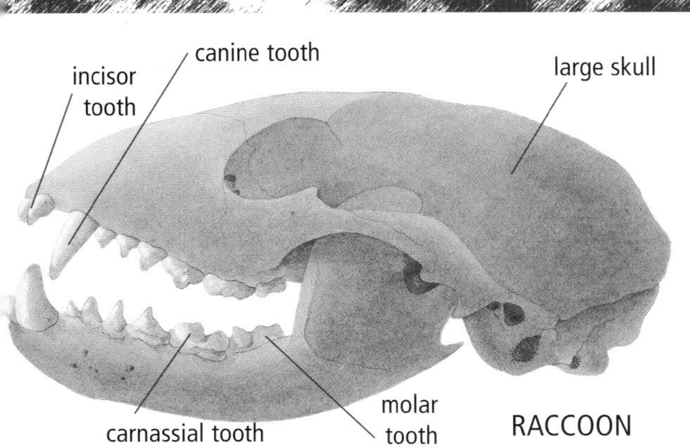

RACCOON

Skulls
Raccoons and their relatives, such as the kinkajou, have a large, tough skull. Raccoons have 40 teeth, while smaller relatives have slightly fewer.

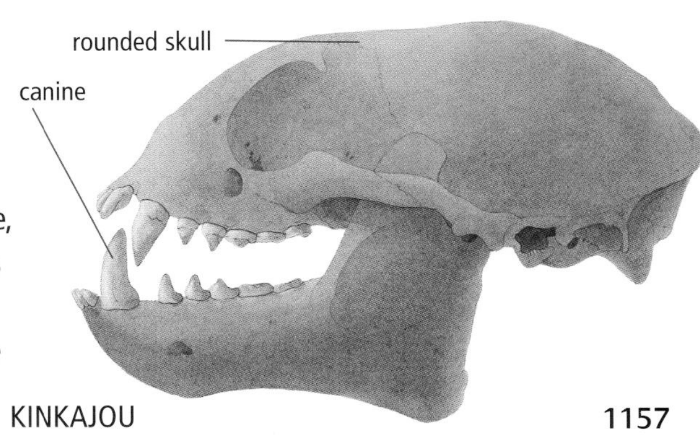

KINKAJOU

From brushland to busy cities

The common raccoon is one of nature's survivors. Most wild animals cannot cope when people alter their habitat. However, far from being reduced in number by the activities of people, raccoons have increased their numbers. They are highly adaptable animals and have changed their lifestyle to survive.

Over the last 100 years, North American raccoons have adapted to living alongside people. As one scientist puts it, "Raccoons take civilization in their stride and grow fat where lampposts have replaced trees."

AMAZING FACTS

- Raccoons live longer in a city than in the countryside.
- City raccoons are smaller than those that live in woodlands.
- The name *raccoon* comes from the Native American word *aroughcun*, which means "the scratcher."
- A raccoon once made a den under the hood of an abandoned car.

Raccoon homes

The common raccoon's natural habitat is woodland or areas of brushland. It also generally lives close to water. However, a raccoon is happy to live anywhere as long as there is plenty of food and places to hide. The raccoon is now just as likely to be living under a house or at the end of the backyard. A favorite raccoon nesting site is in a hollow tree. In the city, raccoons have been known to set up home in the most similar place—a chimney.

The common raccoon lives from southern Canada to

Below: *A baby common raccoon hangs on tightly to a branch using its forepaws. Raccoons can survive just as well in city areas as they do in woodlands.*

Distribution

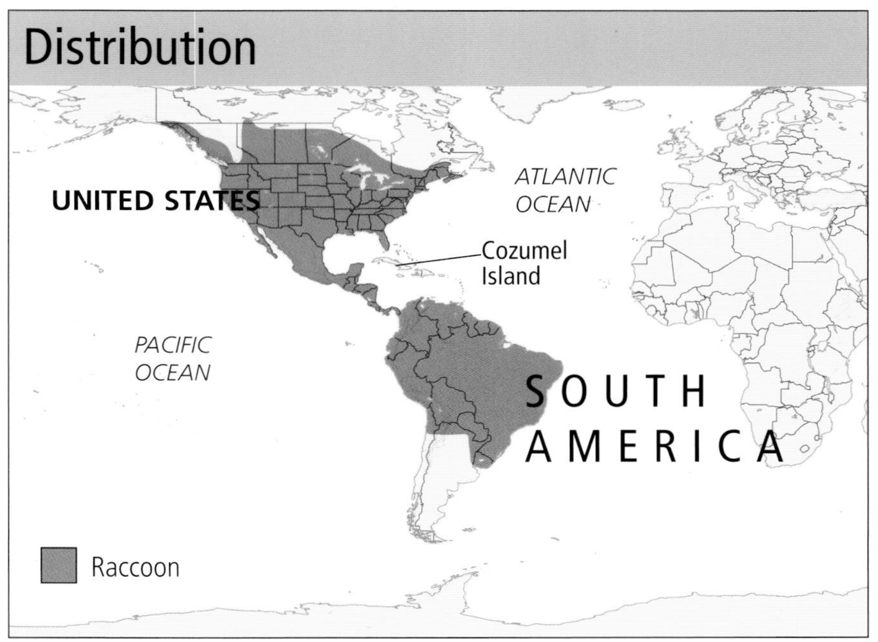

DID YOU KNOW?

Island raccoons

As its name suggests, the Cozumel Island raccoon is now thought to live only on Cozumel Island off the east Yucatán coast of Mexico. This island species looks very similar to the common raccoon, but it is much smaller, weighing only 7 to 9 pounds (3–4 kg), and has a lighter-colored coat compared with the common raccoon.

The Cozumel Island raccoon was probably introduced to the island by people thousands of years ago. The species is classified as endangered because its island home has only a small area of suitable habitat, and it has to compete with people for space.

Central America, where the species gives way to the crab-eating raccoon. This species is similar to the common raccoon, except that it tends to have shorter fur and a longer tail. The crab-eating raccoon lives in the forests and brushlands of South America as far south as Argentina. Most other members of the raccoon family live in forests. The kinkajou, olingos, and coatis live in the rain forests of Central and South America. However, the white-nosed coati also survives in drier habitats as far north as Arizona.

The cacomistle also lives in drier areas, this time in southern Mexico and Central America. Its cousin, the ringtail, lives farther north. This little animal survives in rocky areas across the western United States and is especially common on cliffs.

Climbing to success

Raccoons are nocturnal, although other members of their family, such as coatis, are active during the day. Being nocturnal helps raccoons survive alongside humans because most people are asleep when raccoons come out to feed.

Climbers

Another reason for the raccoon's success is that it never misses an opportunity to find food or a good place to make a den. Raccoons explore every nook and cranny of their environment. To do this, all members of the raccoon family have to be excellent climbers. In the wild, raccoons and coatis are most common on the ground, but they climb into trees to escape their enemies and to rest.

Left: *This raccoon mother has made a nest for herself and her two kits in a warm, snug chimney flue. This is an example of the highly adaptable behavior of common raccoons.*

Above: *After resting in a tree for a few midday hours, a coati forages among the leaf litter to find an insect larva.*

With their prehensile tail, kinkajous are the most arboreal (tree living) of the procyonids. They rarely come down to the ground. However, ringtails are the best climbers. This species does not live in trees. Instead, it makes its home on cliffs. Ringtails can scale even the steepest slopes by clinging on with its claws. Their claws can be withdrawn into the feet to keep them sharp.

Raccoon life

Most raccoon family members live alone. They defend a small territory by marking it with their scent. When two animals meet, they try to frighten each other away with snarls and by bristling their fur to look big and mean.

However, coatis are an exception. Although male coatis live alone, just like their raccoon cousins, female coatis live in family groups, or bands, that can contain up to 25 members. Band members look for food together, sleep together, and take turns to look after the offspring.

DID YOU KNOW?

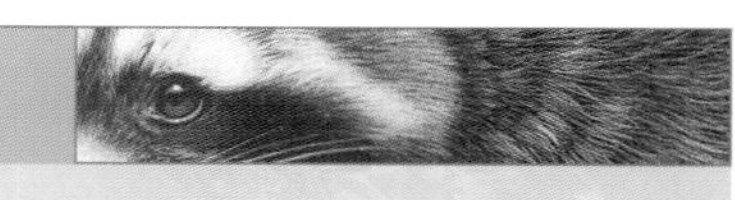

Winter dens

In northern regions, raccoons spend the cold winters in cozy dens. The raccoon sleeps in its den for weeks at a time. However, the animal is not hibernating because its body processes continue more or less as normal. The raccoon wakes up regularly and may leave the den to pass urine or droppings. If the bad weather improves, the raccoon looks for food. Otherwise, it survives by using the body fat it has built up during the fall. A raccoon can lose around 50 percent of its body weight during winter.

Above: *Raccoons never miss the opportunity to scavenge for food, and an overturned garbage can is an excellent target. However, they have such nimble fingers that they are capable of removing the lids themselves if they have to. In addition, they can also open jars and remove corks from bottles. It is no wonder that raccoons have done so well in urban environments.*

DID YOU KNOW?

Noises off

On the rare occasions that raccoons come into contact with one another, they communicate using a large number of noises. In summer and early fall, raccoons produce characteristic hooting calls at dusk. They also purr, whimper, snarl, growl, hiss, scream, and whinny. Most of these noises are intended to scare away other raccoons. However, female coatis snort, grunt, scream, whine, and chatter for more friendly purposes.

Flexible foragers

Although they belong to the Carnivora order, along with Earth's largest and fiercest hunters, raccoons and their relatives do not eat meat alone. The term *carnivore* means "meat eater," but raccoons are more correctly described as omnivores. *Omni* means "all," and an omnivore's diet consists of both plant and animal food.

Active hunters

The ringtail and cacomistle are the most active hunters. They lie in wait for their prey, such as mice and lizards, and then ambush them. Ringtails kill by biting their victims.

The kinkajou eats the least animal food. This tree-living procyonid survives mainly on fruit. It uses it long tongue to lick nectar from flowers. The tongue is also useful for licking honey from bees' nests. This habit has earned the kinkajou the alternative name of honey bear.

Other members of the raccoon family eat a wide range of plant food, including fruit, seeds, and nuts.

They also eat whatever animal food they can find, including small rodents, reptiles, fish, frogs, and insects. Urban and suburban raccoons also dine on the leftover meals of people.

A nose for food

Raccoons find food using their excellent senses of touch and smell. They patrol their territories with their nose close to the ground. Their pointed snout is good for poking into crevices. If there is anything worth investigating inside, the raccoon makes use of its sensitive forepaws.

These paws are essential when looking for food in water. Crab-eating raccoons hunt in water, but like their northern cousin, this species'

DID YOU KNOW?

Washing food

The scientific name for the common raccoon is *Procyon lotor*. The name *Procyon* means "before dog" and it refers to the way raccoons look a little like a hairy fox, which belongs to the dog family Canidae. The name *lotor* means "the washer" in Latin. That name refers to the way raccoons are seen to dip some of their food in water as if washing it. It is unlikely that raccoons are cleaning the food, but nobody knows why they do this.

diet is a mixture of food from all sources.

Coatis forage on the ground in much the same way as raccoons. However, without agile hands, coatis rely on their flexible snout to investigate holes and root into soil for tasty bites to eat.

PREY

Raccoons belong to the Carnivora order of animals. Like other carnivores, they eat mainly other animals. However, raccoons also eat plant foods. Common raccoon prey includes:

LIZARD (order Squamata)
Ringtails, which live in dry, rocky areas, eat a lot of lizards and other reptiles, which are more common in dry places.

FROG (order Anura)
Raccoons often hunt in shallow water. They use their sensitive forepaws to feel for frogs and other aquatic animals at the bottom.

BIRDS AND EGGS (class Aves)
Tree-living coatis often prey on nestlings and birds' eggs.

INSECTS (class Insecta)
All members of the raccoon family eat insects.

FLOWERS AND FRUIT (kingdom Plantae)
Tree-climbing kinkajous eat mainly fruit. They also use their long tongue to lick nectar from the center of large flowers.

Left: *The fruit-eating kinkajou spends the night among the branches, foraging for its favorite food (**1**). Although coatis eat mostly insects and fruit, they will also eat young birds (**2**). Foraging in the shallows of a stream, this common raccoon has fished out its supper (**3**). The most carnivorous of the raccoon family, the ringtail eats spiders, scorpions, snakes, toads, frogs, and in this case, a lizard (**4**).*

Spring into action

Most common raccoons mate between February and March. In the southern part of their range, where it is warmer, they might start breeding as early as December.

At the start of the mating season, males leave their territory and visit their female neighbors. A male raccoon's territory usually overlaps with those of several females, but it never overlaps with a territory belonging to another male.

Nevertheless, males sometimes have to travel far from their home to find mates, and they often meet another male on the way. When this happens, the males try to drive each other away. That behavior rarely results in serious injury, and one male soon backs down.

Each female usually mates with just one male each year. After mating, the male stays with the female for about one week to ensure that no other male mates with her. After a week the male raccoon moves away to find another female.

COMPETITION: Males threaten and fight each other over the right to mate with females. A female raccoon usually mates with only one male each season. The act of mating makes the female's ovaries release their eggs.

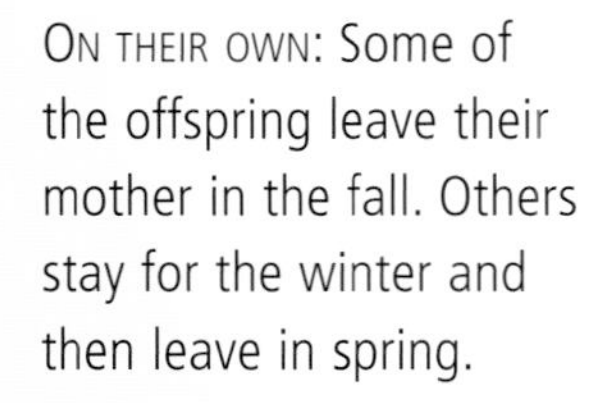

ON THEIR OWN: Some of the offspring leave their mother in the fall. Others stay for the winter and then leave in spring.

A nest for kits

The female is pregnant for about two months. Toward the end of her pregnancy, the female builds a large nest in a hollow tree or another secluded place. She then gives birth to as many as seven kits.

They stay in the nest for 10 weeks, after which they begin to make short trips outside with the mother. The kits eat nothing but solid food after about four months. Other procyonids breed in more or less the same way, although the length of their pregnancy varies by a few weeks. A band of female coatis tends to mate with the same male. He wins a place in the band during the spring, but after he has mated with them, the females chase him away.

TIME TO GROW: After around two months developing inside the mother, the kits are born blind and helpless. However, they develop quickly, and their eyes open after about three weeks.

The life of a raccoon

NIGHT SCHOOL: From the age of 10 weeks, kits leave the den at night and accompany their mother everywhere. She teaches them how to find food.

Common raccoon

MATING SEASON: February to March
GESTATION: 62 days
LITTER SIZE: 3–4; up to 7
WEANED: 4 months
SEXUAL MATURITY: Male 2 years; female 3 years
LIFE SPAN: 13–16 years

Coati

MATING SEASON: February to March
GESTATION: 77 days
LITTER SIZE: 4–6
WEANED: 4 months
SEXUAL MATURITY: Male 3 years; female 2 years
LIFE SPAN: 7 years

AMAZING FACTS

- Unlike other members of the raccoon family, male kinkajous become sexually mature before the females.
- Kinkajous are pregnant for about four months, the longest gestation period of any raccoon.
- Olingo kits do not open their eyes until around 28 days after birth.

Survivors of the fur trade

The "coonskin" hat of a woodsman is a famous symbol of the American frontier. Before European settlers set up home in the forests of North America, the value of a raccoon's warm fur was already well understood by Native Americans. They also hunted raccoons for their meat.

By the seventeenth century, raccoons had joined the beaver, muskrat, and mink as an American animal that was hunted in huge numbers to supply the demand for warm furs in Europe. At the height of the trade in fur at the end of the nineteenth century, more than 1 million raccoons were trapped each year. However, the raccoon is a survivor. Despite huge numbers killed, the species was never in danger.

Apart from the crab-eating raccoon and common raccoon,

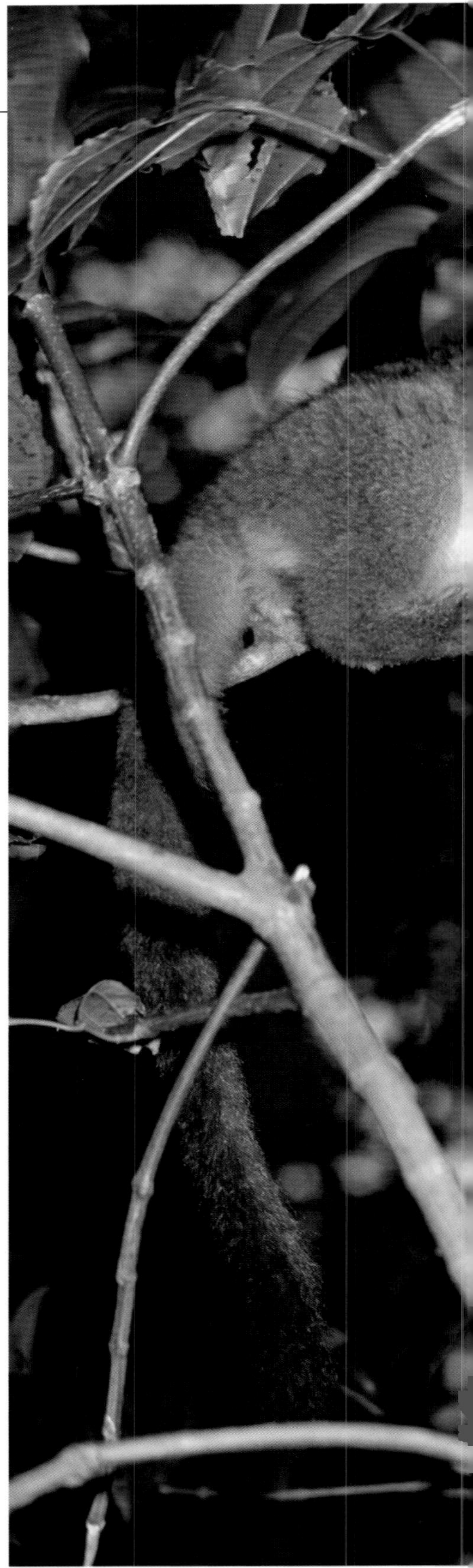

Raccoons in Europe

At risk

This chart shows how the International Union for the Conservation of Nature (IUCN) classifies species and subspecies of raccoons:

Barbados raccoon	*Extinct*
Bahaman raccoon	*Endangered*
Tres Marías Islands raccoon	*Endangered*
Guadeloupe raccoon	*Endangered*
Cozumel Island raccoon	*Endangered*
Cozumel Island coati	*Endangered*
Harris's olingo	*Endangered*

Extinct means that no members of the species or subspecies are living in the wild. *Endangered* means that this species is facing a very high risk of extinction in the wild.

Left: *An olingo perches among the branches in a rain forest. Olingos are threatened by habitat destruction.*

all other species and subspecies of raccoons are threatened with extinction. That is because these raccoons live on islands and have only limited living space. In many cases, the island species are being outcompeted by common raccoons that have been introduced to the islands.

Most other members of the raccoon family are not under threat. However, the cacomistle and two Central American species of olingos are endangered because their habitats are being destroyed.

Raccoons in Europe

Along with the bald eagle, grizzly bear, and rattlesnake, the raccoon is a symbol of the United States. However, raccoons are now becoming more common in Europe.

Raccoons were first brought to Germany in the 1930s to be bred on farms for their fur. Some raccoons were set free, while others escaped from their farms. Since then, the wild population has grown rapidly. European wild raccoons are thought to be descended from animals that escaped when a fur farm was bombed during World War II (1939–1945). Raccoons have spread west and now live in eastern France, Luxembourg, and Belgium.

CHECK THESE OUT

RELATIVES: • African wild dogs • American black bears • Badgers • Bears, small • Brown bears • Cheetahs • Civets • Coyotes • Dholes • Foxes • Hyenas • Jackals • Jaguars • Leopards • Lions • Lynx and wildcats • Marten • Meerkats • Otters • Pandas • Polar bears • Pumas • Red foxes • Tigers • Weasels • Wolverines • Wolves

RAT KANGAROOS

Rat kangaroos are small cousins of the large, more familiar kangaroos that live in Australia. Rat kangaroos are less specialized anatomically than kangaroos. However, they have been around just as long as Australia's most famous big-footed, pouched mammals.

Below: *With their short forelegs and very long hind feet, these rat kangaroos get around by hopping, just as the much larger kangaroos do. As with many rat kangaroos, the tails are nearly as long as the bodies.*

Tiny Hoppers

Everyone knows about kangaroos. They are some of the most unusual and memorable mammals on Earth. Kangaroos have some smaller and less well-known relatives called rat kangaroos. There are 10 species of rat kangaroos, which also go by other names, such as potoroos and bettongs. Rat kangaroos move around by hopping. Like kangaroos, rat kangaroos have long back feet and powerful legs for bounding along.

All in a pouch

Like their larger kangaroo cousins, rat kangaroos live in Australia. They also belong to the group of pouched mammals called marsupials. All mammals are hairy animals that feed their young with milk. The milk is supplied through mammary glands, or teats. Most marsupials have teats inside a pouch, called a marsupium, and young marsupials spend their first weeks and months developing inside this pouch.

Marsupials are extremely undeveloped when they are born. For example, they do not have any back legs, eyes, or fur. They do most of their growing after birth inside the pouch.

Other mammals, such as rats, lions, and humans, are born in a more developed form. The young of these mammals spend a much longer time

KEY FACTS

- **Common name:** Rat kangaroo
- **Scientific name:** Family Potoroidae
- **Species:** 10 species in four genera
- **Range:** Australia and Tasmania
- **Appearance:** Small, pouched marsupials; long hind feet and short forelegs; most have a long tail. All but the musky rat kangaroo have long hind limbs and move quickly by hopping.

- **Common name:** Honey possum
- **Scientific name:** *Tarsipes rostratus*
- **Family:** Tarsipedidae
- **Range:** Southwestern Australia
- **Appearance:** Small, mouselike marsupials; long, thin tongue; narrow, tubular snout; long prehensile (gripping) tail.

Above: *A rufous rat kangaroo forages in leaf litter. These tiny marsupials eat mainly plants.*

RELATIVES

Rat kangaroos and the honey possum are marsupial (pouched) mammals. Most species of marsupials live in Australia and New Guinea. Some species, such as opossums, live in North and South America. Rat kangaroos and honey possums belong to the marsupial order Diprotodontia. This order also includes:

KANGAROO AND WALLABY (family Macropodidae) ▶
These animals are the largest marsupials. They are hopping animals and share many other similarities with rat kangaroos, such long hind feet. Some biologists include the rat kangaroos as a subfamily of the kangaroo family.

KOALA (*Phascolarctos cinereus*) ▶
The koala is one of the best-known marsupials. It lives in the forests of eastern Australia.

POSSUM (order Diprotodontia) ▶
Brushtail possums, pygmy possums, and striped possums make up several families of marsupials that live in Australia, Tasmania, New Zealand, and Melanesia (Vanuatu, New Caledonia, Fiji, and the Solomon Islands). These small animals eat insects and plants.

WOMBAT (family Vombatidae) ▶
Wombats are large, burrowing marsupials that live in the forests, grasslands, and mountains of southern Australia.

Ancestors and relatives

All kangaroos belong to an order of marsupials called Diprotodontia. Members of this group have just two incisors—biting teeth—in the lower jaw. Other types of marsupials have four or more lower incisors.

The first kangaroo-like animals evolved around 30 million years ago. Most of them were small and looked more like rat kangaroos than the larger species now alive.

One unusual rat kangaroo is the musky rat kangaroo. Many biologists think this species is not like other rat kangaroos and put it in a family of its own. For example, musky rat kangaroos are good climbers and walk on all fours rather than with a hopping motion.

Most members of the rat kangaroo family evolved about 15 million years ago. They evolved in open habitats with plenty of grass and few trees.

growing in their mother's uterus. As they grow, the young are nourished by their mother through an organ called the placenta. Marsupials do not have a placenta. Instead, young marsupials are born earlier and much less developed and must be nourished outside the mother's body, inside the pouch.

Right: *A musky rat kangaroo. Apart from the musky rat kangaroo, all the other rat kangaroos have long hind legs and hop. Rat kangaroos live in a wide range of habitats. They are usually larger than possums but smaller and more likely to walk on four legs than wallabies.*

DID YOU KNOW?

The honey possum

The name *possum* is given to several groups of Australian marsupials. The name comes from a Native American word meaning "white dog." This name originally referred to the Virginia opossum, but was shortened to possum and used to describe a range of small marsupials. The honey possum is not closely related to any of the other groups of possums. Like rat kangaroos, it belongs to the order Diprotodontia.

Honey possums are tiny. Their body measures only 3 inches (7.5 cm) long and they have a tail of the same length. This tail is extremely prehensile—it can wrap around objects and grasp them. The honey possum uses its tail as a fifth limb to help it climb among tall grasses and shrubs.

The honey possum also has a long tongue with a brushlike tip. It uses its tongue to lick nectar and pollen from flowers. Other possums eat these types of food along with other foods, but the honey possum survives on nothing but nectar and pollen.

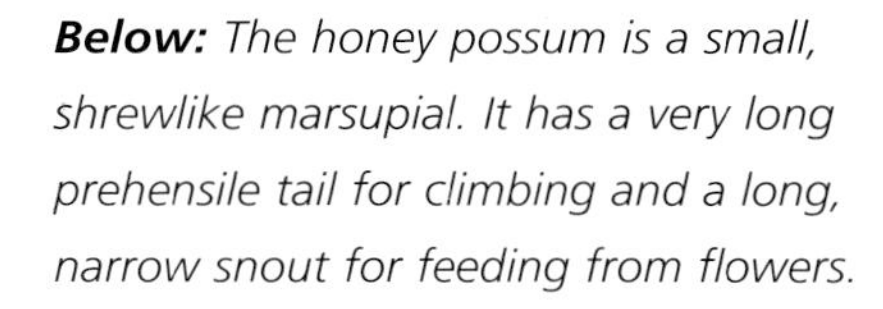

Below: *The honey possum is a small, shrewlike marsupial. It has a very long prehensile tail for climbing and a long, narrow snout for feeding from flowers.*

Before that time, the rat kangaroo's ancestors had lived in trees, but then they became ground-living animals. The long, bounding hind legs were not very useful for climbing, but they were excellent for moving across the ground quickly. Long hind legs also freed up the forepaws for digging and holding food.

The basic kangaroo body form had arrived. All modern species of kangaroos evolved from animals like this.

Left: *Rat kangaroos have very long hind feet, which they use to move quickly by hopping. By comparison, their small forefeet are undeveloped.*

ANATOMY: Rufous rat kangaroo

Coat
The rufous rat kangaroo has coarse red fur. The tips of the fur are gray. That makes the coat appear grizzled.

Eyes
Rat kangaroos are nocturnal, and they have large eyes that see well in the dark.

Forelegs
The rat kangaroo's forelegs are short but strong. When the animal is hopping slowly, it rests on its forepaws. These paws also have long claws for digging up roots.

Hind feet
The long hind feet have four toes. The two middle toes are fused. This thick double toe has a claw that is used for grooming. The two outer toes help the rat kangaroo balance.

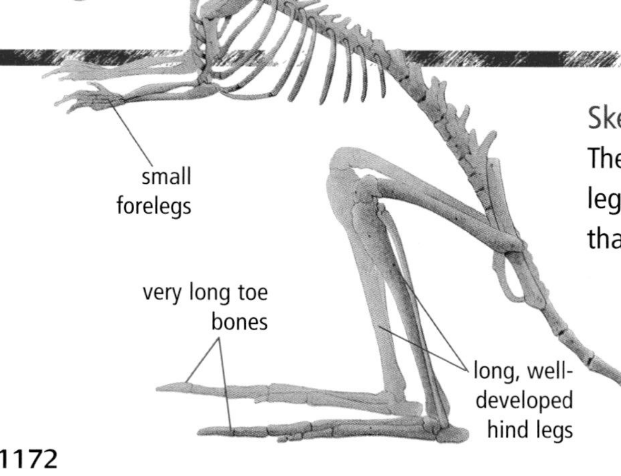

Skeleton
The rat kangaroo's hind legs are much larger than its forelegs.

Teeth
Rat kangaroos have unusual premolar teeth. These long, blade-shaped cheek teeth are used to cut up food before it is ground up by the flatter molar teeth behind.

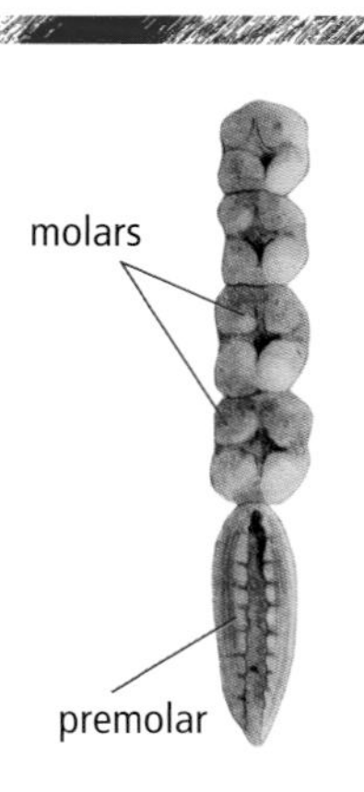

Bettong feet

The burrowing bettong is the only big-footed member of the family known to live in burrows on a regular basis. The long-clawed forefeet are used for burrowing and digging up food, while the long hind feet move the animal along quickly.

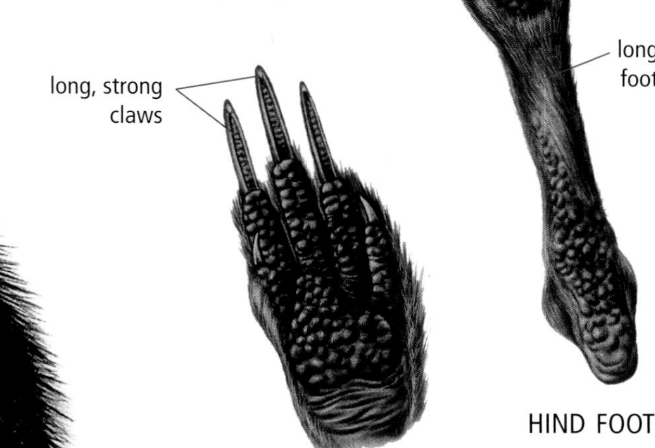

HIND FOOT

FOREFOOT

FACT FILE

The largest kangaroo rat is the burrowing bettong (right), which is 15.5 inches (39.5 cm) long. Its tail is 12 inches (30 cm) long. The single species of honey possum (above, far right) is much smaller. At just 2.5 to 3.3 inches (6.4–8.4 cm) long, it is one of the smallest marsupials. Its tail is generally slightly longer than its body.

Rufous rat kangaroo

GENUS: *Aepyprymnus*
SPECIES: *rufescens*

SIZE

HEAD–BODY LENGTH: 15 inches (38 cm)
TAIL LENGTH: 13.5 inches (34 cm)
WEIGHT: 7 pounds (3.2 kg)

COLORATION

Red-brown fur with white undersides; fur is grizzled—the base of the hairs is darker than the tips.

Tail

The tail is thick and muscular. When the animal is resting, it is propped up by the tail. When the rat kangaroo moves, the tail is held above the ground and acts as a counterbalance.

Skulls

Rat kangaroos have a wide head, a narrow snout, and a strong jaw, which is used to crush a range of foods. Honey possums have a very long snout, which makes the jaw weak. Honey possums do not need to chew, however; they lick up their food.

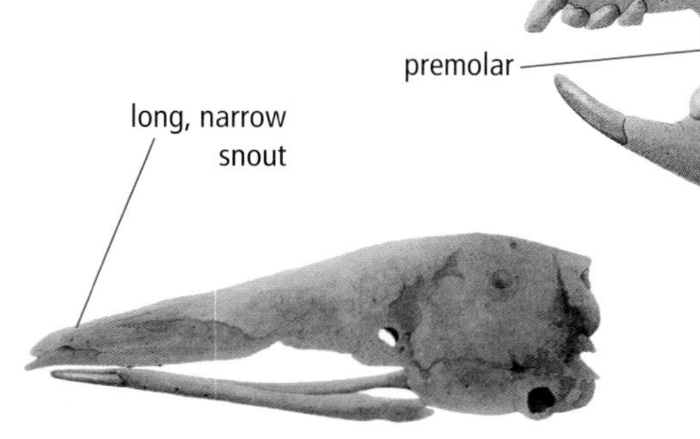

HONEY POSSUM

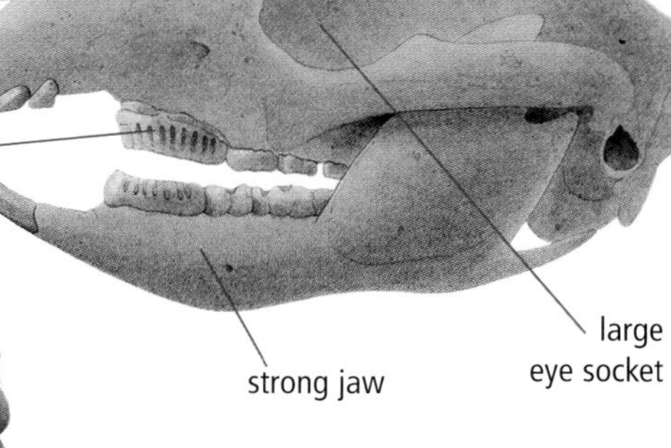

RAT KANGAROO

Outback animals

The central region of Australia is a hot desert. Most rat kangaroos live closer to the coast, where there is enough rain for plants to grow for most of the year. The musky rat kangaroo lives in the rain forests that grow in the state of Queensland in the northeast of Australia. This part of Australia receives huge amounts of rain, and there is enough water in the soil to supply the trees of a forest.

Less rain falls farther south and in western Australia, so there is less moisture available for plants. As a result, only

DID YOU KNOW?

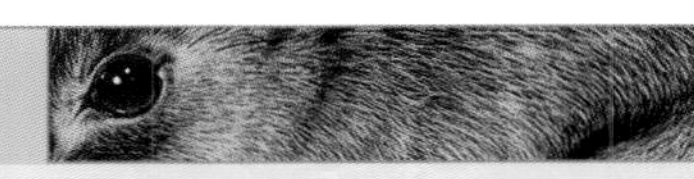

Helped by fire

Wild fires are very destructive. Not only do they destroy people's homes but they also burn down trees and kill wildlife. However, the brush-tailed bettong relies on fires to survive. Australian grass and shrublands need fires every few years. The fire stimulates seeds to grow and removes dead wood. The fire also encourages the growth of mushrooms and fungi—the favorite food of the brush-tailed bettong. The rat kangaroo runs ahead of the flames before ducking into a disused burrow. If it cannot find somewhere to hide, it turns and leaps through the flames to the burned land behind.

Right: *A musky rat kangaroo eats an insect. Its sharp teeth easily cut through tough insect coverings. These rat kangaroos have shorter hind legs than other species and do not hop.*

Distribution

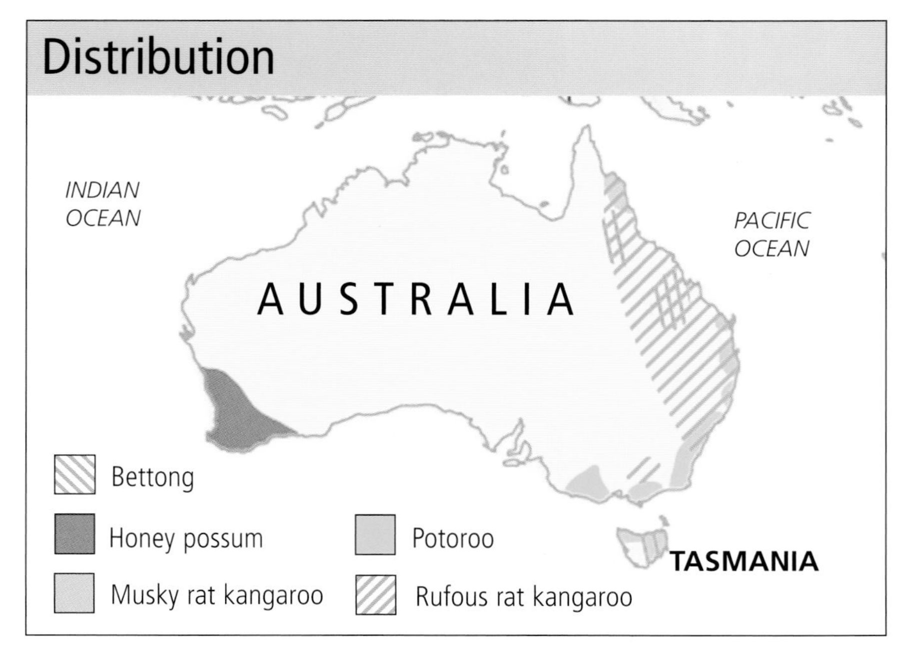

a few trees grow there; most plants are small and fast-growing. The habitats in these regions are usually grasslands,

> BY THE 1940S, BURROWING BETTONGS SURVIVED ON JUST THREE ISLANDS OFF THE COAST OF WESTERN AUSTRALIA.

bushlands, or in some places woodlands of eucalyptus trees. This area is often called the outback, which is where most rat kangaroos live.

At risk

This chart shows how the International Union for the Conservation of Nature (IUCN) classifies rat kangaroos:

BURROWING BETTONG, OR BOODIE	*Vulnerable*
NORTHERN BETTONG	*Endangered*
GILBERT'S POTOROO	*Critically endangered*
LONG-FOOTED POTOROO	*Endangered*
BROAD-FACED POTOROO	*Extinct*
DESERT RAT KANGAROO	*Extinct*

Extinct means that no members of the species are living in the wild. *Critically endangered* means that this species is facing an extremely high risk of extinction in the wild. *Endangered* means that this species is facing a very high risk of extinction in the wild. *Vulnerable* means that this species is facing a high risk of extinction in the wild.

The three species of potoroos live in the dry woodlands and brush areas of eastern and southern Australia. (A fourth species is extinct.) They inhabit areas only where there is thick vegetation covering the ground, in which they can stay hidden. The rufous rat kangaroo and most species of bettongs live wherever tall grasses grow.

DID YOU KNOW?

Burrowing bettong

When European explorers arrived in Australia, the burrowing bettong, or boodie, was one of the most common animals. Other rat kangaroos were restricted to areas where there was plenty of thick ground cover to hide in. However, burrowing bettongs lived in many habitats across the south and west of the country, from semideserts to woodlands. However, by the 1940s these kangaroo rats survived on just three islands off the coast of Western Australia. They had been outcompeted for space and food on the mainland by another burrowing animal—the rabbit—which had been introduced to Australia 90 years earlier.

Silent night moves

Rat kangaroos are rarely seen in the wild. All but one species live alone, and most rat kangaroos never go out during the day. The honey possum is also nocturnal, and because it is so tiny, it is also difficult to spot.

Since they live alone and are active at night, rat kangaroos and honey possums do not make many noises; nor do they communicate using visual signals.

Day and night

Just because people do not see or hear rat kangaroos, it does not mean that they are not busy. Most of these animals spend the day in dome-shaped nests built of leaves, bark, and twigs. Each nest has a single entrance, and the nest is often built in a shallow dip dug into the ground. Burrowing bettongs dig complicated warren systems in the soil.

After night falls, rat kangaroos come out from their nests and begin to search for food. Long-nosed potoroos

***Below:** Burrowing bettongs often dig their warrens among boulders. They also sometimes make burrows in the floors of caves. They prefer light, crumbly soil but can also manage to dig in firmly packed ground.*

often start feeding as dusk falls, and in some places they come out even earlier.

Once again, the musky rat kangaroo is an exception; it feeds in the early morning and late evening, before returning to the nest at night.

DID YOU KNOW?

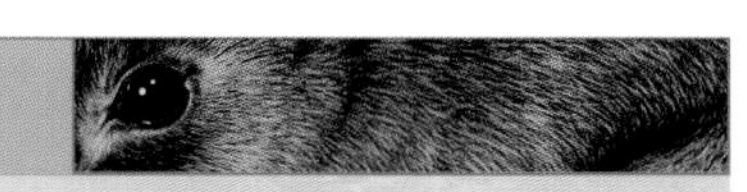

Flexible tail tips

Rat kangaroos are descended from tree-living animals. These ancestors had long, prehensile tails that they used as a fifth limb when climbing—just as honey possums now do. Modern species of rat kangaroos cannot wrap their tails around objects in the same way. However, the tips of their tails are still flexible, and rat kangaroos use them for another purpose. Rat kangaroos line their nests with grass and shredded leaves. To do so, they make a bundle of this bedding material and then pick it up by curling their tail around it so they can carry it home. Once in the nest, the kangaroo rat lifts up the old bedding with its snout and pushes the new bedding underneath.

Claiming territory

Rat kangaroos look for food inside a certain area, or home range. A home range contains a few nesting sites, and a rat kangaroo stays at each place for about one month before moving on. Each rat kangaroo's home range overlaps with those of its neighbors, but the animals rarely cross into each other's territory.

One species of rat kangaroo—the burrowing bettong—lives in colonies. Each colony can contain more than 100 individuals living in networks of burrows similar to those made by European rabbits. Smaller groups of burrowing bettongs are more common, however. The simplest groups contain a single male and a harem (group) of several females, all of which live in the same burrow. In certain places, these harems live in the same area to make up the larger colonies.

Anything goes for omnivores

Rat kangaroos are omnivores and eat whatever food is available. An omnivore is an animal that eats both animal and plant food. The term *omnivore* means "all eater." The plant food eaten by rat kangaroos includes fruit, seeds, flowers, leaves, stems, bulbs, and roots. Fungi, such as mushrooms, are also a common part of their diet. Most of the animals eaten by rat kangaroos are invertebrates (animals without a backbone), such as insects and earthworms.

DID YOU KNOW?

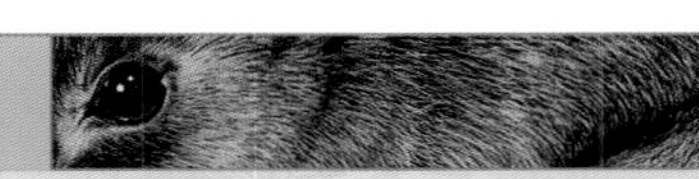

Probing powers

Honey possums have been known to climb into trees to find blossoms, but most of their feeding is done among the grass and shrubs of Australia's heathlands in the southwest of the country. With large, grasping feet and a long, flexible tail, the honey possum is extremely agile. It can climb straight up stalks and even walk upside down under a branch. Honey possums weigh only 0.4 ounce (12 g) and they can reach flowers at the end of slender twigs without breaking or bending the twigs too much.

The possum's long, brush-tipped tongue collects droplets of sweet nectar and grains of pollen. Once the possum has put the food into its mouth, it scrapes the food off its tongue by rubbing it against its teeth and several hard ridges on the roof of its mouth. Wild honey possums can survive on flower food. The nectar is a good source of water and sugar, while pollen contains proteins and essential minerals.

Meat on the menu

The musky rat kangaroo eats more animal food than other rat kangaroo species. It hunts for prey on the forest floor and along the low branches of trees. It grabs prey food in its mouth and then holds it in its forepaws while eating the meal. Many other rat kangaroos do not actively hunt insects. Instead, they swallow tiny animals living on the marsupials' plant foods.

Despite its hunting habits, the musky rat kangaroo's diet is made up mostly of fallen fruit, which it

Below: *A potoroo enjoys a mushroom meal (***1***). This bettong is nibbling the leaves on a low-growing plant. These animals also eat roots and tubers (***2***). Insects are an important food source for the musky rat kangaroo (***3***).*

picks up from the ground. Other rat kangaroos forage in the same way, looking for seeds and fruit on the ground. They eat only leaves and the tough parts of a plant if there is nothing else available.

Rat kangaroos also dig for food using their powerful forearms and long claws. They dig up roots, tubers (small, fleshy, underground stems), and fungi. Rat kangaroos can also find insects to eat by digging.

COMPETITORS

Except for a few species, such as the red kangaroo, most marsupials have been seriously affected by nonmarsupial mammals introduced to Australia. Rat kangaroos are no exception. They have been reduced in number by introduced predators, such as cats and foxes. Other animals do not kill rat kangaroos, but they compete with them for food and habitats. These competitors include:

LORY (subfamily Loriinae)
These small parrots feed on nectar and pollen just as honey possums do. These birds also have a long tongue with a brushlike tip. Australian lories feed mainly in trees and are too big to perch on grass. The honey possum survives because it is small enough to climb through grass stems to reach food.

RABBIT (family Leporidae)
Rabbits were introduced to Australia about 150 years ago. In just 100 years, 24 rabbits bred into a population of 600 million. Rabbits compete with rat kangaroos for food, and they have also taken away living space from the burrowing bettong.

Left: *Two honey possums eat nectar and grains of pollen from banksia flowers. They use their narrow, pointed snout to probe into the blossoms and a long, rough-ended tongue to lick up the sweet nectar and pollen grains.*

Pouch hangers-on

Rat kangaroos can breed from the age of about one year. Most species breed throughout the year. The rufous rat kanagroo produces a single offspring, while the musky rat kangaroo gives birth to two or three offspring. At birth, marsupial offspring are hairless and wormlike. A rat kangaroo's offspring is 0.6 inch (15 mm) long. It drags itself through a forest of fur across the mother's belly to the pouch. Once inside, the offspring immediately begins to suck milk from a teat. It does not let go for two months. During this period, the offspring grows at a rapid rate. After two months, it makes short trips out of the pouch before returning for more milk. At about four months, the young rat kangaroo is too large for the pouch, and the mother refuses to let it back in. By this time, the pouch probably holds a younger brother or sister.

Pausing development

Female rat kangaroos mate just a few hours after giving birth. However, their body is not ready to become pregnant again. In addition, the offspring that has just been born will be in the pouch for at least the next two months. Therefore, the development of the next embryo is delayed. This baby starts growing again after a few weeks so that it is born just as the previous baby leaves the pouch. If a baby growing in the pouch dies for some reason, the next embryo begins to develop immediately to make use of the empty pouch.

Rufous rat kangaroo

Mating season: Throughout the year
Gestation: 22–24 days
Litter size: 1
Number of litters: 3 per year
Weight at birth: 0.03 ounce (1 g)
Leaves pouch: 16 weeks
Weaned: 22 weeks
Sexual maturity: 11–13 months
Life span: 6 years

Musky rat kangaroo

Mating season: February to July
Gestation: Not known
Litter size: 2 or 3
Number of litters: 1 per year
Weight at birth: 0.03 ounce (1 g)
Leaves pouch: 21 weeks
Weaned: 22 weeks
Sexual maturity: 12 months
Life span: Not known

On their way: At about two months of age, the offspring leave the pouch and ride on their mother's back as she looks for food. Honey possums are fully grown and sexually mature at six months.

LEAVING THE POUCH: The young rat kangaroo stays in the pouch for two months before making short trips outside. By the age of four months, the young marsupial leaves the pouch for good.

The life of a rat kangaroo

TINY OFFSPRING: The tiny offspring is born after just few weeks. It is barely visible as it crawls through the mother's fur from the birth canal to the pouch. The offspring attaches itself to the teat inside the pouch, where it continues its development. The mother often mates again soon after giving birth.

CHECK THESE OUT

RELATIVES: • Kangaroos • Koalas • Opossums • Possums • Wombats

WINTER MATINGS: Honey possums generally mate in winter when food is easier to find. Females mate with several males.

The life of a honey possum

INSIDE THE POUCH: The litter contains about four offspring, each of which might have a different father. As they suckle, the offspring increase five times in weight before leaving the pouch.

AMAZING FACTS

- The sperm cells of a male honey possum are the largest of any animal in the world.
- Female honey possums are larger than the males.
- Honey possum offspring are the smallest of any mammal. They weigh just 0.00002 ounce (0.0005 g).

RATS AND MICE

Rats and mice are small, gnawing mammals, or rodents. Rats and mice are extremely successful creatures and live virtually worldwide, from south of the Arctic Circle to Australia and South America.

Below: *A whistling rat eats some twigs and leaves. These Old World rodents live in dry areas of southwestern South Africa and make shrill whistling sounds to communicate.*

KEY FACTS

- **COMMON NAME:** Rats and mice
- **SPECIES:** More than 1,000 species in around 230 genera
- **SUBFAMILIES:** Most species of rats and mice are placed in Murinae (Old World rats and mice) and Sigmodontinae (New World rats and mice). There are eight other subfamilies of Old World rats and mice.
- **HABITAT:** All habitats except tundra and ice
- **RANGE:** Worldwide
- **APPEARANCE:** Small rodents with a flexible body, sharp teeth, sensitive whiskers, and a long tail; medium to dark brown coat on back and flanks, lighter on the underside; some species have stripes.

Gnawing Success

Rats and their smaller relatives, mice, are the most familiar rodents of all. People are most used to seeing the house mouse and the sewer rat, which is more correctly called the brown rat. However, there are more than 1,000 species of rats and mice in the world.

Toothy tools

Rats and mice are rodents. Rodents form the largest order of mammals. Almost half of all mammal species are rodents. Other rodents include squirrels, marmots, beavers, guinea pigs, chinchillas, and porcupines. The term *rodent* comes from the Latin word *rodere*, meaning "to gnaw," which is exactly what rodents do. They use their long front teeth to gnaw food, dig tunnels, and cut up nesting material.

Mice and rats form the largest single subgroup of rodents, grouped together in the family Muridae. This family also contains several familiar types of rodents, such as lemmings and hamsters.

In simple terms, mice and rats are split into two groups: Old World rats and mice and New World rats and mice. Old World species live in Africa, Europe, Asia, and Australia. New World rats and mice are

Above: *A wood mouse forages among the leaf litter. These mice have large ears and eyes, a long tail, and dark brown fur with white underparts.*

RELATIVES

Rats and mice belong to the order Rodentia and the family Muridae. They are the largest and most widespread group of rodents. Rats and mice are split into two main subfamilies—Sigmodontinae and Murinae. Other subfamilies in the family Muridae include:

GERBIL (subfamily Gerbillinae) ▶
Gerbils are desert rodents that move around by hopping on long hind legs. They live in Africa and western and southern Asia.

HAMSTER (subfamily Cricetinae) ▶
Hamsters are mouselike rodents that live in dry parts of eastern Europe and Asia. The golden hamster, a rare species from Syria, is the wild ancestor of pet hamsters.

BLIND MOLE RAT (subfamily Spalacinae)
Sometimes placed in a separate family, these tunneling rodents live in southeastern Europe and the Middle East. Their eyes are completely covered by skin and fur, and they do not have outer ears or a tail. Blind moles rats dig using giant front teeth, and build elaborate tunnel networks.

VOLE ▶ **AND LEMMING** ▼ (subfamily Arvicolinae)
As well as lemmings and voles, this subfamily of rodents also includes the muskrat. Voles and lemmings live across the northern hemisphere and are especially common in cold areas and near to water.

American species. Although both groups of rodents contain animals that look the same and live in similar ways, they are not closely related (see the box opposite).

Types of rats and mice

Rats and mice are so successful because they all have the same type of small and flexible body, sharp teeth, sensitive whiskers, and a long tail. This simple body form allows them to thrive in so many places around the world. However, the similar body form also makes it difficult to tell apart the huge number of different species of rats and mice.

There are nearly 600 Old World species. They include field mice, harvest mice, rock rats, water rats, bandicoot rats,

Above: *There are around 80 species of South American rice rats, ranging from Central America to Argentina.*

Below: *The leaf-eared mouse of South America is named for its large, protruding ears.*

Above: *Vesper rats live high in the trees and have large, clawed toes on their hind legs to help them climb.*

DID YOU KNOW?

Ancestors

New World and Old World rats and mice are both large groups of animals. At first glance many of the species from the two groups look very similar and live in very similar ways and places. However, biologists know that the two groups are not all that closely related. New World rats and mice, for example, are more closely linked to lemmings and hamsters. One way to distinguish New World species from Old World species is to look at differences in their cheek teeth, or molars.

The reason why many of the unrelated species look so similar is convergent evolution. This process occurs when two species that live in the same habitat but in different parts of the world evolve in the same way. For example, the deer mouse, which lives in North America, is very similar to the long-tailed field mouse, which lives across Europe and Asia. Both species have evolved to survive among grass and brushes and to eat the same food. Therefore, evolution has altered their body form into the same shape to suit the way they live.

groove-toothed swamp rats, and many more. The smallest species is the African pygmy mouse, which weighs just 0.2 ounce (6 g). The largest is Cuming's slender-tailed cloud rat from the Philippines. This species is 20 inches (50 cm) long.

There are around 430 New World rats and mice. Most of them live in South America. They include deer mice—the most common species in North America—rabbit rats, vesper rats, leaf-eared mice, swamp rats, and climbing rats.

Above: *The pygmy mouse is the smallest New World rodent. This seed eater lives in grassy nests, often beneath stones.*

Above: *The 12 species of burrowing mice live across most of tropical South America (Peru, Bolivia, and Brazil).*

Left: *The house mouse (far left) and the brown rat (left) are among the most common of all the Old World rodents.* **Above:** *The fish-eating rat lives in northwestern South America.*

ANATOMY: Brown rat

Muzzles

Rats are really large versions of mice, although house mice have a more pointed muzzle and their ears are further forward. New World mice and rats, such as the spiny rice rat, vesper rat, South American field mouse, and the Central American water mouse, have a similar shape.

Touch-sensitive skin

A rat has very sensitive skin on the muzzle and long whiskers that can detect the slightest movements and air currents. These sensors, combined with a good sense of smell, allow rats to move around in total darkness.

Ears

Rats have a good sense of hearing. However, compared with mice, a rat's ears are small relative to its head.

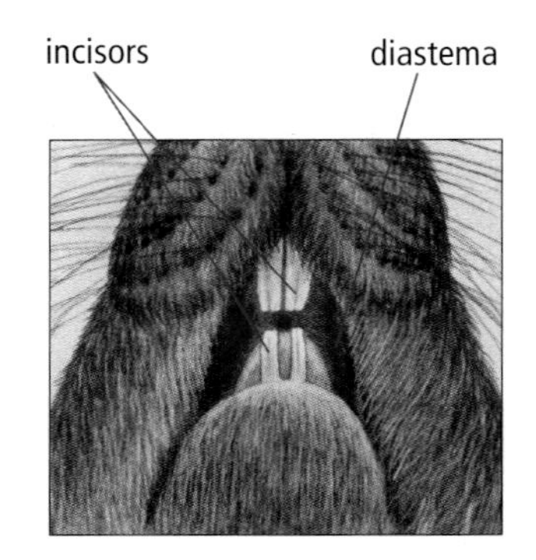

Teeth

Rodents are gnawing animals, and rats are no exception. The long incisors (biting teeth) grow continuously, so they never wear away. There is a gap called the diastema between the incisors and the other teeth. That gap allows the animal to gnaw on objects from the side as well as from the front.

Front paws

Both forepaws have four toes, each bearing a sharp nail.

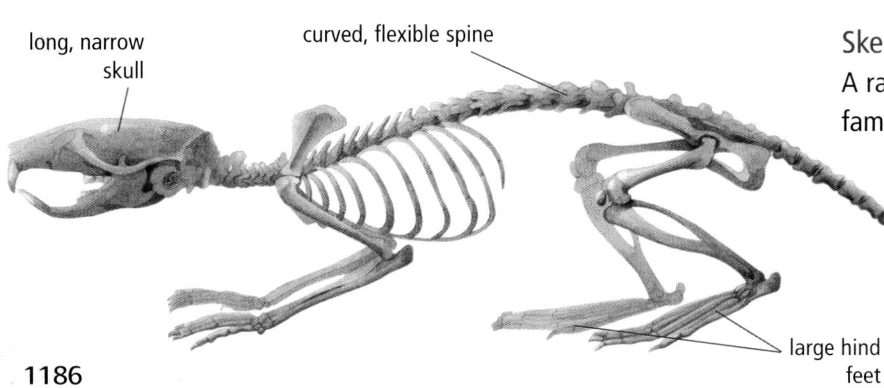

Skeleton

A rat's skeleton is typical of a member of the family Muridae. Most Old and New World species look similar. The skeleton is light and flexible and allows the rat to run jump, swim, stand on its hind legs, and squirm through gaps.

Paw prints

The paw prints of a mouse (right) and a rat (far right) show the differences between the forepaws, which each have four toes, and the hind paws, which each have five toes.

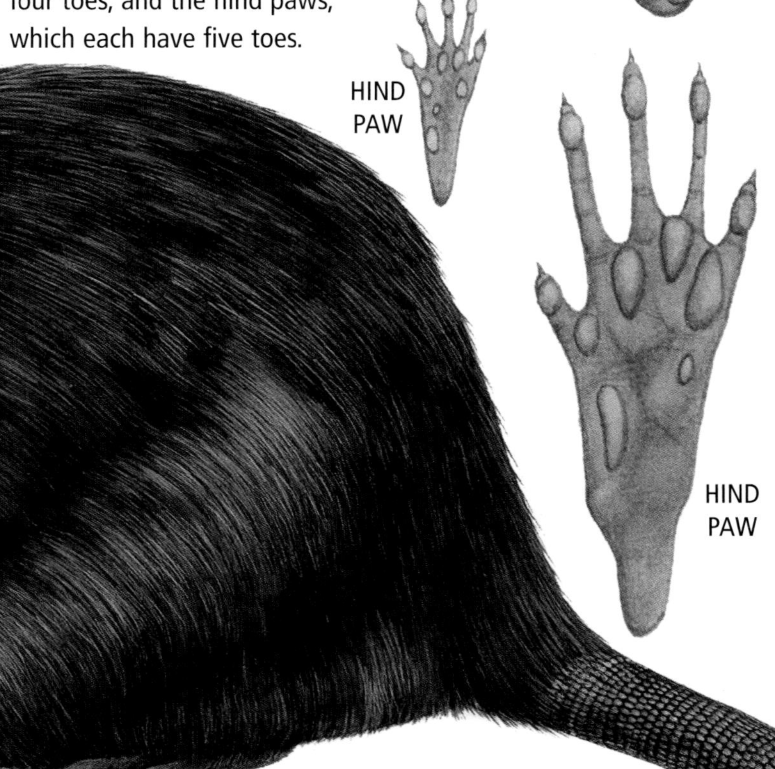

Hind foot

The hind feet are larger than the forepaws. Each hind foot has five toes.

Tail

The long tail is hairless and flexible. The rat uses it as a feeler and to balance its body when running.

FACT FILE

The brown rat (above left) is around 12 inches (30 cm) long from nose to tail. The house mouse (above right) is about one-third this size, between 3.2 and 3.6 inches (8–9 cm) long. The South American giant rat is slightly larger than brown rats. The smallest New World mouse is the pygmy mouse, measuring 3.5 inches (9 cm) from nose to tail.

Brown rat

GENUS: RATTUS
SPECIES: NORVEGICUS

SIZE

HEAD–BODY LENGTH: 9–10.6 inches (23–27 cm)
TAIL LENGTH: 7–8.7 inches (18–22 cm)
WEIGHT: 7–14 ounces (200–400 g); males are around 40 percent heavier than females

COLORATION

Gray-brown to black. Many rats bred in captivity are yellow.

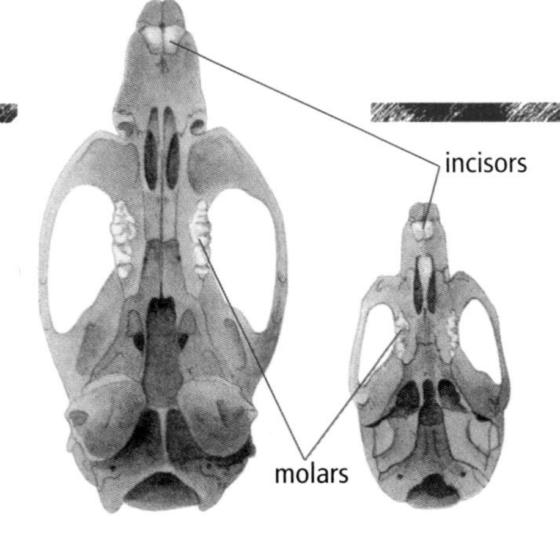

BROWN RAT MOUSE

Short, sharp teeth

A rodent's incisor teeth are long and chisel-shaped. They have long, deep roots, which make the teeth very strong. The teeth grow continuously but are kept sharp by the animal's constant gnawing and biting.

BROWN RAT

diastema

incisors

molars

Skull

The brown rat's skull clearly shows the gnawing, ever-growing incisors common to all rodents. The molars are used to grind and chew food. There is a large gap, called the diastema, between the incisors and molars.

All around the world

Most New World rats and mice live in South America, but they only migrated there around five million years ago.

South America was cut off from North America and the rest of the world for millions of years until about five million years ago. During this period of isolation, hamsterlike rodents were evolving into the first species of New World rats and mice. They evolved while living alongside other mammals, such as cottontail rabbits and moles. These rats and mice evolved to avoid competing with their nonrodent neighbors.

From north to south

When North America joined onto South America, other rats and mice migrated into the southern continent.

> NEW WORLD MICE AND RATS HAVE EVOLVED TO LIVE IN ALL NONMARINE HABITATS.

Few mammals lived in South America at that time, and there were no rabbits, moles, or other mammals common in North America. New World rats and mice evolved quickly into new species that lived in habitats just like those occupied by nonrodents in North America. For example, rabbit rats lived among thickets and swamps just as cottontails do; shrew mice were small and hunted in thick grass; and mole mice were expert tunnelers.

New World mice and rats now live from the tip of Cape Horn to the edge of the Arctic tundra. They have evolved to live in all nonmarine habitats in between, from the tops of rocky mountains, to tropical forests and parched deserts. The most widespread species are rice rats and deer mice.

Distribution

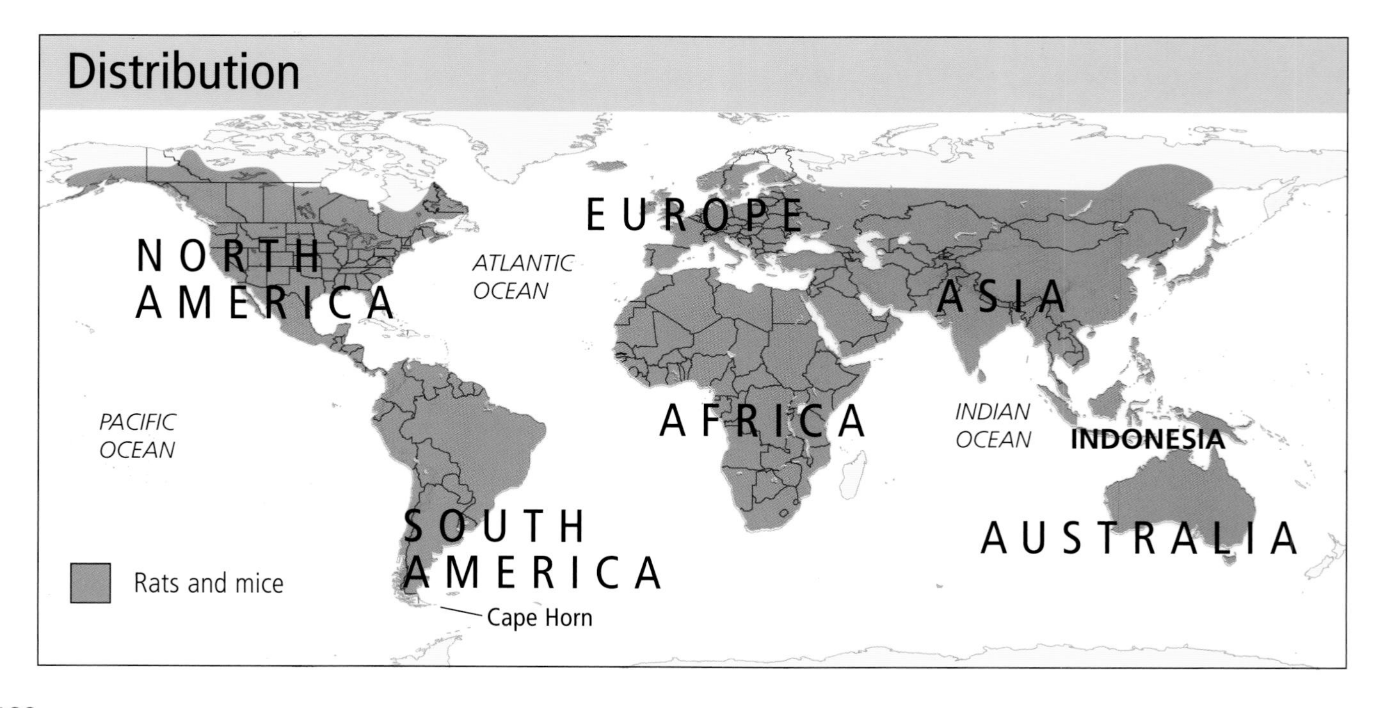

AMAZING FACTS

- The marmoset rat of Southeast Asia lives only in bamboo forests. It makes its nest inside hollow bamboo stems.
- The Brazilian shrew mouse is a tunneling mammal in South America. It lives underground and feeds on insects.
- The hopping mice of Australia have large ears filled with blood vessels. The blood in the ears is cooled by the air to help the mice lose heat.
- The house mouse is the only mammal to live permanently in Antarctica.

Human influence

Most of the many Old World rat and mouse species live in specific local areas. For example, the Flores long-nosed rat lives on just one island in Indonesia. Just as the New World species do, Old World species are adapted to live in all types of habitats. However, three Old World species have adapted to life alongside people. As a result, these rodents—the house mouse, the brown rat, and the black rat—have spread around the world.

Above: *A brown rat emerges from a drain cover. These rats eat practically anything and live in their millions, particularly in urban areas.*

The black rat is originally from India and lives only in warmer parts of the world. The brown rat is thought to be originally from northern China but now lives throughout the world. The house mouse also spread from Asia and is now the most widespread mammal on Earth after people.

Sensing danger

Rats and mice are nocturnal animals and rarely come out during the day. They generally live alone and spend the day in nests or burrows. Burrowing species include grasshopper rats of the southwestern United States. They dig U-shaped tunnels that are always around the same size—20 inches (50 cm) long and 5.5 inches (14 cm) below the ground. Nest-building species include wood rats and deer mice.

Above: *A wood mouse spends most of the winter asleep in its underground burrow, waking now and then to feed on its acorn store.*

Below: *A pack rat brings a sprig of cactus to place at the entrance to its den in a desert in the southwestern United States. The spines prevent would-be intruders from entering the rat's den.*

Once out in the open, rats and mice use their senses to find food and stay out of danger. They have large eyes, but sight is their weakest sense. For much of the time, these rodents move around in near darkness and must rely on their senses of smell, hearing, and touch.

DESPITE THEIR SIZE, RATS AND MICE ARE RELATIVELY INTELLIGENT AND THEY CAN REMEMBER COMPLEX ROUTES.

Climbers and swimmers

Rats and mice are successful because they can find food almost anywhere. Their sharp, gnawing teeth are useful for tackling most types of food, but they have to find it first. They do that by being inquisitive and checking out every corner of their habitat for signs of food.

Most rats and mice are good swimmers, and their strong and flexible body also makes them excellent climbers.

Despite their size, rats and mice are also relatively intelligent and they can remember complex routes.

AMAZING FACTS

- Wood mice cannot see red light very well, so biologists can follow them at night using a flashlight with a red filter.
- Marsh rats are expert swimmers, so they often build nests of woven grass in reeds surrounded by water.
- Deer mice do not clean their nests. Once the nest is dirty, the mice move out and make another nest somewhere else.

DID YOU KNOW?

Collect callers

Pack rats (below) are one of nature's collectors. These rodents from the western United States and Mexico make their nests in crevices. They usually use twigs and leaves to line their nests and place pebbles and sticks to fill in gaps in the walls. They protect their homes with cactus spines. However, pack rats take almost anything they can find to use as building materials, and they are especially fond of shiny objects, such as foil wrappers, coins, and even small spoons and other cutlery. The shiny objects may scare predators away from the nest.

Pack rats will drop whatever they are carrying to trade it for a shiny item. As a result, people often find a pebble or nutshell in place of something shiny. This habit has earned pack rats the alternative name of trader rats.

Everything on the menu

The diets of rats and mice are made up mainly of plant food, such as seeds, leaves, and fruit. These rodents can adapt to living almost anywhere. The key to the rodents' success is being able to eat whatever is available to survive.

The teeth of rats and mice are always sharp and ready to gnaw into all types of food. In wild habitats, mice and rats often eat insects, spiders, and other minibeasts as well as their plant diet. Larger species, such as rats, also kill frogs and small lizards. Those rodents that live among people feed on a different type of food. The rats that live in the world's sewers, for example, eat anything from soap and glue to vegetable peelings and bones. The other thing that makes rodents so successful is their ability to seek out food. They are extremely curious animals that search everywhere for food. They poke their sensitive nose into nooks and crannies in search of the whiff of a meal.

Right: *A South American climbing rat has found a berry to eat. These rats eat a wide range of food, including nuts and leaves.*

Below: *A grasshopper mouse closes in on its prey (**1**). A fish-eating rat lives up to its name. These New World rodents feel for aquatic prey using their sensory whiskers (**2**). Although usually a plant eater, this hungry cotton rat has caught a crayfish (**3**).*

Their touch-sensitive whiskers and tails allow the rodents to forage for food even in total darkness. With their flexible and agile bodies, rats and mice can climb, leap, or wriggle into almost any spot. In all, these feeding skills make rats and mice nature's most effective mammal species.

Above: *A wood mouse nibbles on a beetle while balancing and hiding among the branches of a thorny bush.*

PREDATORS

Rats and mice can be victims of most kinds of predators because they are generally fairly defenseless against attack. Few rats and mice survive for more than two years, but they breed as quickly as possible before they are killed or die. Predators of rats and mice include:

CAT (family Felidae) ▶
The domestic cat was bred from the wild cat to hunt the mice that were living in people's homes. Other cat species that hunt for small animals, such as rats and mice, include bobcats and servals.

BIRD OF PREY (class Aves) ▶
Eagles, hawks, and other birds of prey are a constant threat to rats and mice. These birds have the most powerful sense of sight of any animal and can see rats and mice moving through grass from high up in the sky. Owls hunt at night and track their prey by listening for the faint sounds they make.

SNAKE (order Squamata) ▶
Snakes either squeeze rats and mice to death or poison them with venom. Many rodent-hunting snakes have heat-sensitive pits on their head. These pits allow the snakes to sense the body heat of the rats and mice in the dark in the same way that an infrared camera works.

Above: *A brown rat has caught a large frog. This is an important source of protein for the rat, which must eat one-third of its body weight each day. Rats are very adaptable and will eat almost anything, including bones, soap, and glue.*

Short, busy lives

Life is short for rats and mice. If they are not killed by predators, they are still in constant risk from their environment. Because they are small creatures, rats and mice are unable to survive long periods of cold or drought, and they can only go for a few hours without food before they begin to starve. Unlike other types of rodents, such as dormice, rats and mice do not hibernate. A few species, such as deer mice and wood mice, go into a deep sleep during cold periods and other times when food is hard

HIDDEN AWAY: The newborn mice are hidden in a sturdy, spherical nest with a small, single opening. This drawing is a cross section through the nest to reveal the baby mice inside.

The life of a harvest mouse

TIME TO GO: At 15 days the young mice leave the nest for good. Their mother is now pregnant with the next litter.

Harvest mouse

GESTATION: 19 days
LITTER SIZE: 4–6
NUMBER OF LITTERS PER YEAR: 2–4
WEIGHT AT BIRTH: 0.025 ounce (0.7 g)
FIRST SOLID FOOD: 15 days
WEANED: 15–16 days
SEXUAL MATURITY: 6–8 weeks
LIFE SPAN: 18 months

Deer mouse

GESTATION: 21–40 days
LITTER SIZE: 3–4
NUMBER OF LITTERS PER YEAR: 3–5
WEIGHT AT BIRTH: 0.08 ounce (2.3 g)
FIRST SOLID FOOD: 13 days
WEANED: 3–4 weeks
SEXUAL MATURITY: 7 weeks
LIFE SPAN: 2 years

to find. However, unlike hibernating animals, these mice sleep for only a few days at a time and often wake up to feed on stored food.

Rats and mice can breed from an early age, and they breed often. This behavior compensates for them not living a long life. In this way, they make sure that some of their genes survive on in their offspring. Most species produce large litters and can have several litters each year. In the right conditions, many rats and mice can easily produce 20 offspring in a year. Young rats and mice are able to breed after just a few months.

MOVING WITH MOTHER: At six days old the offspring have learned to lie still while their mother carries them around in her mouth.

EXPLORATION TIME: At 12 days old, the young mice are able to see and move around the nest. They may also go outside for short trips.

Self-sharpening teeth

Rats and mice are gnawing animals. They do not simply gnaw on food, however; they use their teeth to cut nesting materials and for digging and tunneling. They need to have sharp, biting teeth throughout their life in order to survive. Like all rodents, rats and mice have self-sharpening biting teeth that never wear away or become blunt. These teeth have hard enamel on the front surface and a softer material called dentine covering the inner surface. The teeth grow continuously, and the dentine is constantly worn away as the animal bites its teeth together.

Island threats

With the huge numbers of house mice and rats living across the world, it is hard to imagine that most species of rats and mice are very rare. Rare species have usually evolved to survive in only a certain part of the world. For example, many rat and mouse species live only on islands. The ancestors of these rodents probably floated to the islands by chance on logs and branches. After many years on the islands, the rodents evolved into new species.

However, many of these rare species are now in danger of extinction. People living on the islands have destroyed the rodents' natural habitats and introduced new animals to the area. These animals include other rodents, such as house mice and rats, which probably arrived by accident with the people on their boats.

Pet animals, such as cats, are also a threat to rare rodent species. The island rats and mice have probably lived for millions of years without being hunted by large predators, and they are easy pickings for introduced hunters.

MANY RARE SPECIES OF RATS AND MICE ARE NOW IN DANGER OF EXTINCTION.

Galápagos Islands

This map shows the position of the Galápagos Islands.

Galápagos rice rats

The Galápagos Islands off the coast of Ecuador are home to many unusual animals. The islands' wildlife was studied by English naturalist Charles Darwin (1809–1882). Darwin was interested to see how his theory of evolution could explain how animals had evolved while cut off from the mainland.

The islands are home to several species of rice rats that do not live anywhere else in the world. These already-rare rats have now become very

At risk

This chart shows how the International Union for the Conservation of Nature (IUCN) classifies rats and mice:

LONG-FOOTED WATER RAT	*Critically endangered*
GIANT THICKET RAT	*Endangered*
FLORIDA MOUSE	*Vulnerable*

Critically endangered means that this species is facing an extremely high risk of extinction in the wild. *Endangered* means that this species is facing a very high risk of extinction in the wild. *Vulnerable* means that this species is facing a high risk of extinction in the wild.

Above: *This mouse has fallen prey to a cat. Pet cats are responsible for killing a huge number of rats and mice, some species of which are now rare.*

rare indeed. People have destroyed their habitats and introduced cats and other rats to the islands. At least five of the islands' rice rat species have become extinct.

CHECK THESE OUT

RELATIVES: • Beavers • Cabybaras and coypus • Ground squirrels • Hamsters and dormice • Mole rats • Squirrels • Voles and lemmings

DID YOU KNOW?

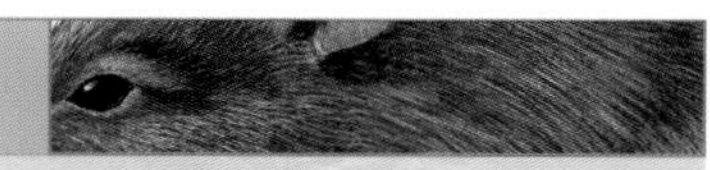

Carrying disease

Rats and house mice often spread diseases, which is why people try to keep them out of their houses. Perhaps the most deadly rodent is the black rat, which spread the bubonic plague across the world. The bubonic plague, or Black Death, killed 100 million people in the sixth century and 75 million in the fourteenth century. The disease is caused by a bacterium (single-celled microorganism) that infects fleas that live on black rats. When these fleas bite people, the fleas pass on the disease to them.

Other diseases spread by rats and mice include Weil's disease and toxoplasmosis. Weil's disease is a dangerous flulike illness caused by a bacterium in rat's urine. People can catch the disease if dirty water gets into cuts. Toxoplasmosis is caused by a parasite. Most people are unaffected by it, but the disease can be harmful to newborn babies.

FAMILY TREE

This family tree reveals the relationships among mammals based on physical and genetic (inherited) similarities. All mammals are members of the class Mammalia. Within this class, mammals are organized into several large groups called orders and then into smaller families. Within an order, for example Carnivora, more closely related mammals are placed into a particular family, such as Felidae (the cat family). The branches of the tree show how closely related particular groups are to each other. Volume and page numbers refer to articles in this encyclopedia.

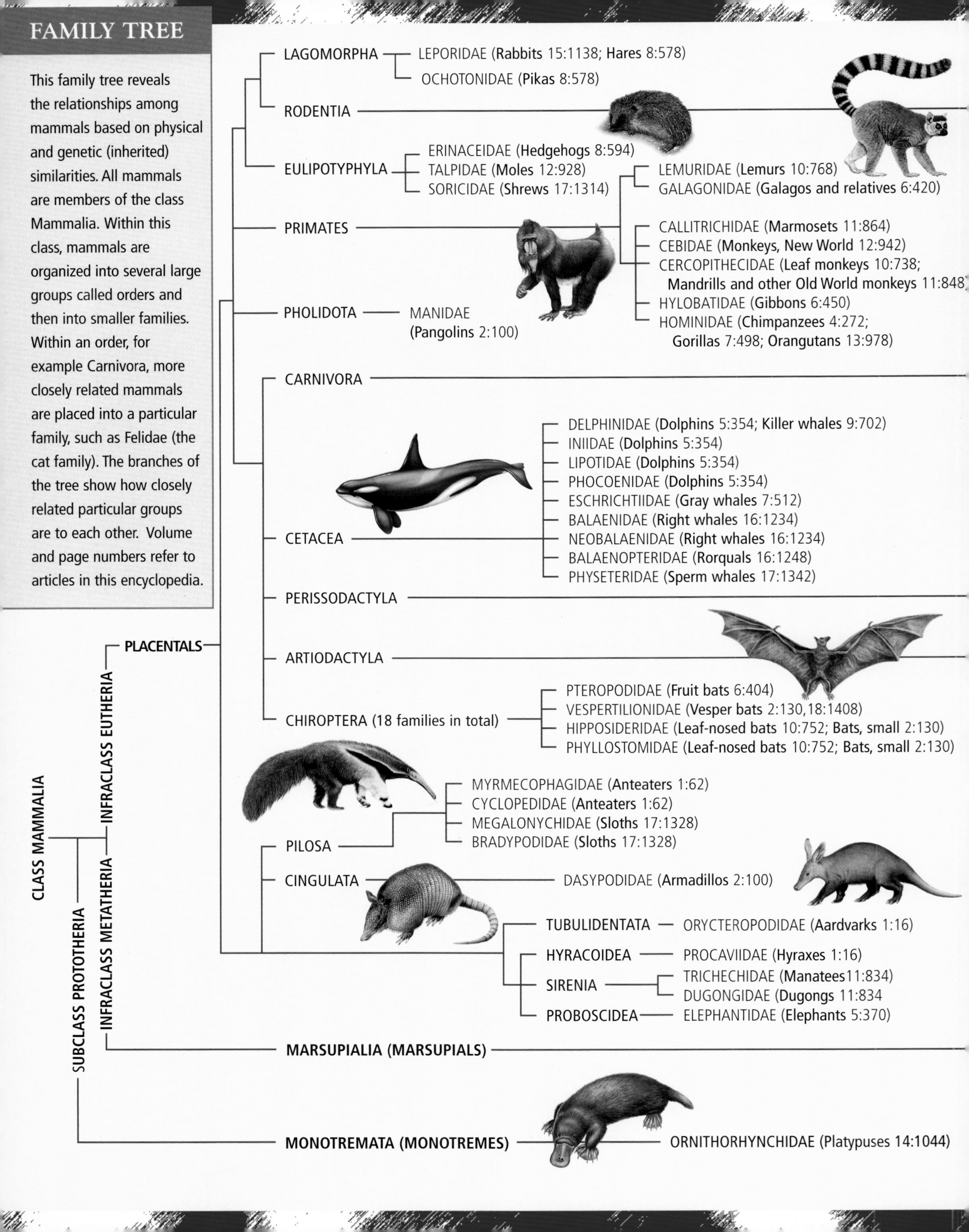

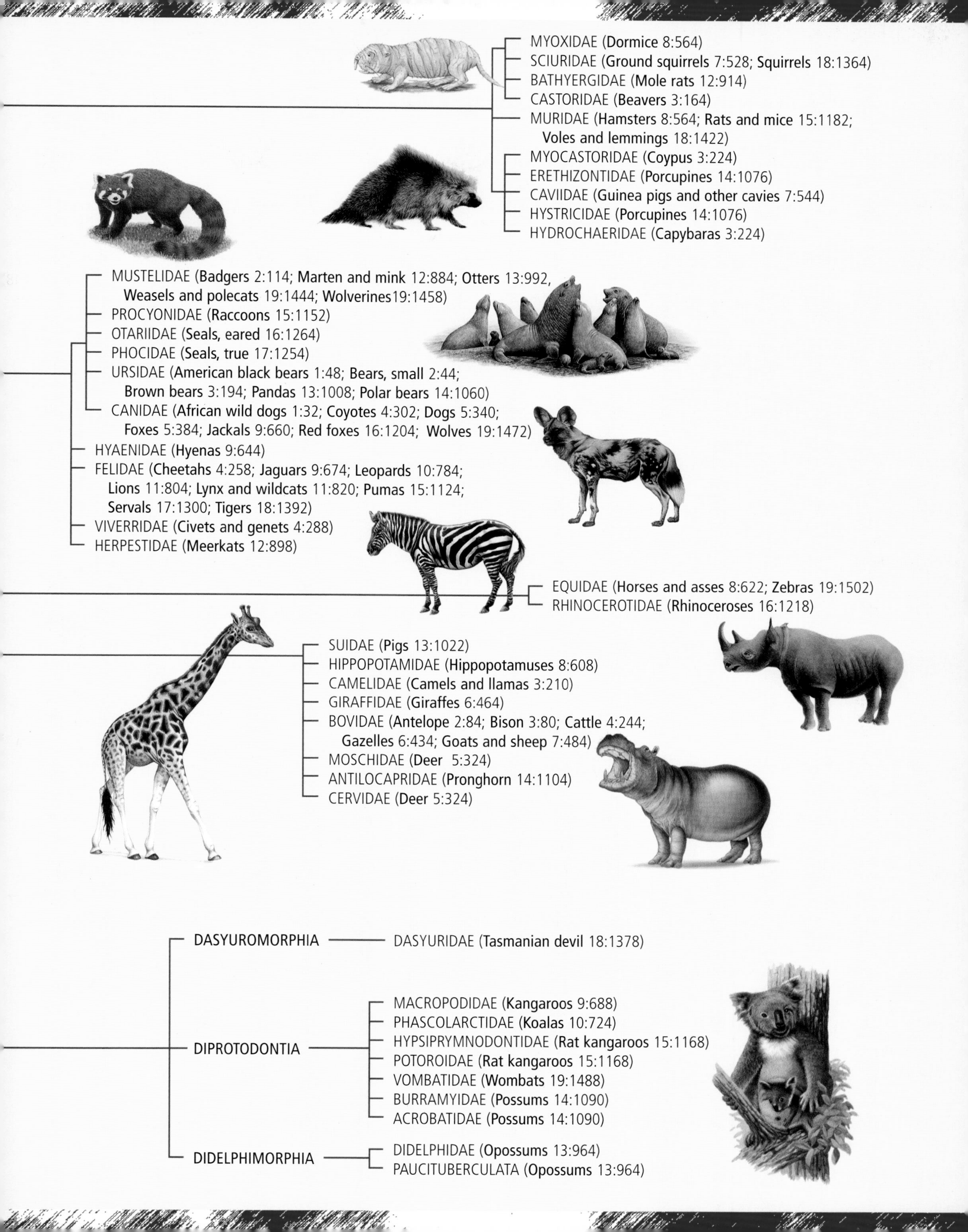

MYOXIDAE (Dormice 8:564)
SCIURIDAE (Ground squirrels 7:528; Squirrels 18:1364)
BATHYERGIDAE (Mole rats 12:914)
CASTORIDAE (Beavers 3:164)
MURIDAE (Hamsters 8:564; Rats and mice 15:1182; Voles and lemmings 18:1422)
MYOCASTORIDAE (Coypus 3:224)
ERETHIZONTIDAE (Porcupines 14:1076)
CAVIIDAE (Guinea pigs and other cavies 7:544)
HYSTRICIDAE (Porcupines 14:1076)
HYDROCHAERIDAE (Capybaras 3:224)
MUSTELIDAE (Badgers 2:114; Marten and mink 12:884; Otters 13:992, Weasels and polecats 19:1444; Wolverines19:1458)
PROCYONIDAE (Raccoons 15:1152)
OTARIIDAE (Seals, eared 16:1264)
PHOCIDAE (Seals, true 17:1254)
URSIDAE (American black bears 1:48; Bears, small 2:44; Brown bears 3:194; Pandas 13:1008; Polar bears 14:1060)
CANIDAE (African wild dogs 1:32; Coyotes 4:302; Dogs 5:340; Foxes 5:384; Jackals 9:660; Red foxes 16:1204; Wolves 19:1472)
HYAENIDAE (Hyenas 9:644)
FELIDAE (Cheetahs 4:258; Jaguars 9:674; Leopards 10:784; Lions 11:804; Lynx and wildcats 11:820; Pumas 15:1124; Servals 17:1300; Tigers 18:1392)
VIVERRIDAE (Civets and genets 4:288)
HERPESTIDAE (Meerkats 12:898)
EQUIDAE (Horses and asses 8:622; Zebras 19:1502)
RHINOCEROTIDAE (Rhinoceroses 16:1218)
SUIDAE (Pigs 13:1022)
HIPPOPOTAMIDAE (Hippopotamuses 8:608)
CAMELIDAE (Camels and llamas 3:210)
GIRAFFIDAE (Giraffes 6:464)
BOVIDAE (Antelope 2:84; Bison 3:80; Cattle 4:244; Gazelles 6:434; Goats and sheep 7:484)
MOSCHIDAE (Deer 5:324)
ANTILOCAPRIDAE (Pronghorn 14:1104)
CERVIDAE (Deer 5:324)
DASYUROMORPHIA
DASYURIDAE (Tasmanian devil 18:1378)
DIPROTODONTIA
MACROPODIDAE (Kangaroos 9:688)
PHASCOLARCTIDAE (Koalas 10:724)
HYPSIPRYMNODONTIDAE (Rat kangaroos 15:1168)
POTOROIDAE (Rat kangaroos 15:1168)
VOMBATIDAE (Wombats 19:1488)
BURRAMYIDAE (Possums 14:1090)
ACROBATIDAE (Possums 14:1090)
DIDELPHIMORPHIA
DIDELPHIDAE (Opossums 13:964)
PAUCITUBERCULATA (Opossums 13:964)

Index

Page numbers in *italic type* refer to illustrations. Page numbers in **boldface** refer to main articles on a subject.

A
Ambush 1162

B
Bettong *1174*, *1178*
Burrowing (boodie) *1173*, 1175, *1176*, 1177
Bobcat 1126, 1130–1131, *1134*, 1135, 1148
Burrowing
burrowing bettong *1176*
rabbit *1146*, *1147*
rats and mice 1190

C
Cacomistle 1153, 1159, 1162
Calling, raccoon 1161
Canines, cat 1135
Carnivores 1154
Cat
Big and small cats 1126
Civet *see* Ringtail
Geoffroy's 1127, 1131
Mountain 1127, 1130, 1131
Otter *see* Jaguarundi
Pampas 1126, 1127
Tiger 1127, 1131
Cellulose digestion 1148
Coati *1153*, *1155*, 1159, 1160, 1161, 1163, *1163*, 1165
Ring-tailed 1157
Cottontail *1141*, 1146
Desert 1144
New England 1145
Cougar 1125

D
Digestion, double *1149*
Diseases, carried by rats and mice 1197

E
Ermine (stoat) 1148
Evolution
cats 1127
kangaroo 1171
lagomorphs 1140
rabbit 1140
rat kangaroo 1170–1171
rats and mice 1188
Evolution, convergent, rats and mice 1185

F
Fighting for dominance, rabbit 1150

G
Gerbil 1184

H
Hamster 1184
Hare 1140–1141, *1143*, *1144*
Harems, burrowing bettong 1177
Hunted by man, raccoon 1166
Hyoid bone 1126

I
Incisors
lagomorph 1140
rabbit *1143*
rat *1186*, *1187*
rodent 1140
Introduced species
and rat kangaroos 1179
rats and mice 1196–1197

J
Jackrabbit 1140
Jaguarundi (otter cat) 1126, 1127, 1131, 1132, *1133*

K
Kinkajou (honey bear) 1153, *1154*, *1157*, 1159, 1160, 1162, *1163*, 1165
Kodkod 1126, 1127, 1129, 1130, 1131

L
Lagomorphs 1139–1141
Lemming 1184
Living in groups
burrowing bettong 1177
coati 1161
rabbit 1146–1147
Lory 1179
Lynx 1129, 1135, 1137
Canadian 1126, 1130

M
Margay (tigrillo) 1127, 1131, 1132, *1135*
Marsupials (pouched mammals) 1170
Mole rat, Blind 1184
Mouse **1182–1197**
Brazilian shrew 1189
Deer 1185, 1190, 1191, 1194–1195
Harvest *1194–1195*
Hopping 1189
House *1185*, 1187, 1189
New World 1183–1184, 1185, 1188–1189
Old World 1183, 1184–1185, 1188
Pygmy *1185*, 1187
Wood *1183*, *1190*, 1191, *1193*, 1194–1195
Mustelids 1154
Myxamatosis 1144–1145

N
Nocturnal
cats 1132–1133
honey possum 1176
puma *1132*
raccoon 1160
rat kangaroo 1176
rats and mice 1190

O
Ocelot 1126, 1127, 1131, 1132
Olingo 1153, 1154, 1159, 1165
Omnivores
raccoon 1162–1163
rat kangaroo 1178–1179

P
Panda, Red *1152*, 1155
Panther, Florida 1130, 1133
Parental care
mouse *1194–1195*
puma 1136, *1137*
rabbit *1151*
rat kangaroo *1181*
Pika 1140, 1143
Plague, bubonic 1197
Possum, Honey 1169, *1171*, *1173*, *1174*, 1176, *1179*, *1180–1181*
Potoroo *1174*, *1178*
Long-nosed 1176–1177
Procyonids 1153–1155
Puma (mountain lion; cougar; panther) **1124–1137**

R
Rabbit **1138–1151**, 1179
Bunyoro 1146
Domestic 1141
European *1139*, *1141*, *1142*, 1144, *1146*, 1150
Pygmy 1141, 1143, 1146
Riverine 1145
Volcano 1144, 1145
Raccoon **1152–1167**
Common *1152–1160*, *1163*–1167
Cozumel Island 1154, 1159, 1167
Crab-eating 1154, 1159, 1162–1163
Rat **1182–1197**
Black 1189
Brown *1186–1187*, *1189*, *1193*
Cuming's slender tailed cloud 1185
Fish-eating *1185*
Grasshopper 1190
Marmoset 1189
Marsh 1191
New World 1183–1184, 1185, 1188–1189
Old World *1182*, 1183, 1184–1185, 1188
Pack (trader rat) *1190*
Rice *1184*
South American giant 1187
Vesper *1184*
Whistling *1182*
Rat kangaroo **1168–1181**
Musky *1170*, *1174*, 1177, *1178*–1179, 1180
Rufous *1169*, *1172–1173*, *1174*, 1175, 1180
Ringtail (civet cat) 1153, *1155*, 1159, 1160, 1162, *1163*
Rodents 1139–1140
teeth 1195

S
Scent marking
rabbit 1146–1147, *1150*
raccoon 1160
Stalking, by cats 1135

T
Tail, prehensile, kinkajou *1154*, 1160
Tigrillo *see* Margay
Toxoplasmosis 1197

V
Vole 1184

W
Washing food, raccoon 1162
Weil's disease 1197